ESSENTIALS OF GAME THEORY

Synthesis Lectures on Artificial Intelligence and Machine Learning

Intelligent Autonomous Robotics
Peter Stone
2007

A Concise Introduction to Multiagent Systems and Distributed Artificial Intelligence
Nikos Vlassis
2007

Essentials of Game Theory: A Concise, Multidisciplinary Introduction
Kevin Leyton-Brown and Yoav Shoham
2008

Essentials of Game Theory

Kevin Leyton-Brown and Yoav Shoham

www.morganclaypool.com

ISBN: 9781598295931 paper
ISBN: 9781598295948 ebook

DOI: 10.2200/S00108ED1V01Y200802AIM003

A Publication in the Morgan & Claypool Publishers series
SYNTHESIS LECTURES ON ARTIFICIAL INTELLIGENCE AND MACHINE LEARNING #3

Lecture #3
Series Editor: Ronald J. Brachman, Yahoo! Research and Tom Dietterich, Oregon State University

Library of Congress Cataloging-in-Publication Data

Series ISSN: 1939-4608 print
Series ISSN: 1939-4616 electronic

ESSENTIALS OF GAME THEORY

A Concise, Multidisciplinary Introduction

Kevin Leyton-Brown
University of British Columbia, Vancouver, BC, Canada
http://cs.ubc.ca/~kevinlb

Yoav Shoham
Stanford University, Palo Alto, CA, USA
http://cs.stanford.edu/~shoham

M&C MORGAN & CLAYPOOL PUBLISHERS

ABSTRACT

Game theory is the mathematical study of interaction among independent, self-interested agents. The audience for game theory has grown dramatically in recent years, and now spans disciplines as diverse as political science, biology, psychology, economics, linguistics, sociology and computer science–among others. What has been missing is a relatively short introduction to the field covering the common basis that anyone with a professional interest in game theory is likely to require. Such a text would minimize notation, ruthlessly focus on essentials, and yet not sacrifice rigor. This Synthesis Lecture aims to fill this gap by providing a concise and accessible introduction to the field. It covers the main classes of games, their representations, and the main concepts used to analyze them.

"This introduction is just what a growing multidisciplinary audience needs: it is concise, authoritative, up to date, and clear on the important conceptual issues."

—Robert Stalnaker, MIT, Linguistics and Philosophy

"I wish I'd had a comprehensive, clear and rigorous introduction to the essentials of game theory in under one hundred pages when I was starting out."

—David Parkes, Harvard University, Computer Science

"Beside being concise and rigorous, Essentials of Game Theory is also quite comprehensive. It includes the formulations used in most applications in engineering and the social sciences, and illustrates the concepts with relevant examples."

—Robert Wilson, Stanford University, Graduate School of Business

"Best short introduction to game theory I have seen! I wish it was available when I started being interested in the field!"

—Silvio Micali, MIT, Computer Science

"Although written by computer scientists, this book serves as a sophisticated introduction to the main concepts and results of game theory from which other scientists, including social scientists, can greatly benefit. In eighty pages, Essentials of Game Theory formally defines key concepts, illustrated with apt examples, in both cooperative and noncooperative game theory."

—Steven Brams, New York University, Political Science

"This book will appeal to readers who do not necessarily hail from economics, and who want a quick grasp of the fascinating field of game theory. The main categories of games are introduced in a lucid way and the relevant concepts are clearly defined, with the underlying intuitions always provided."

—Krzysztof Apt, University of Amsterdam, Institute for Logic, Language and Computation

"To a large extent, modern behavioral ecology and behavioral economics are studied in the framework of game theory. Students and faculty alike will find this concise, rigorous and clear introduction to the main ideas in game theory immensely valuable."

—Marcus Feldman, Stanford University, Biology

"This unique book is today the best short technical introduction to game theory. Accessible to a broad audience, it will prove invaluable in artificial intelligence, more generally in computer science, and indeed beyond."

—Moshe Tennenholtz, Technion, Industrial Engineering and Management

"Excerpted from a much-anticipated, cross-disciplinary book on multiagent systems, this terse, incisive and transparent book is the ideal introduction to the key concepts and methods of game theory for researchers in several fields, including artificial intelligence, networking, and algorithms."

—Vijay Vazirani, Georgia Institute of Technology, Computer Science

"The authors admirably achieve their aim of providing a scientist or engineer with the essentials of game theory in a text that is rigorous, readable and concise."

—Frank Kelly, University of Cambridge, Statistical Laboratory

KEYWORDS

Game theory, multiagent systems, competition, coordination, Prisoner's Dilemma: zero-sum games, Nash equilibrium, extensive form, repeated games, stochastic games, Bayesian games, coalitional games

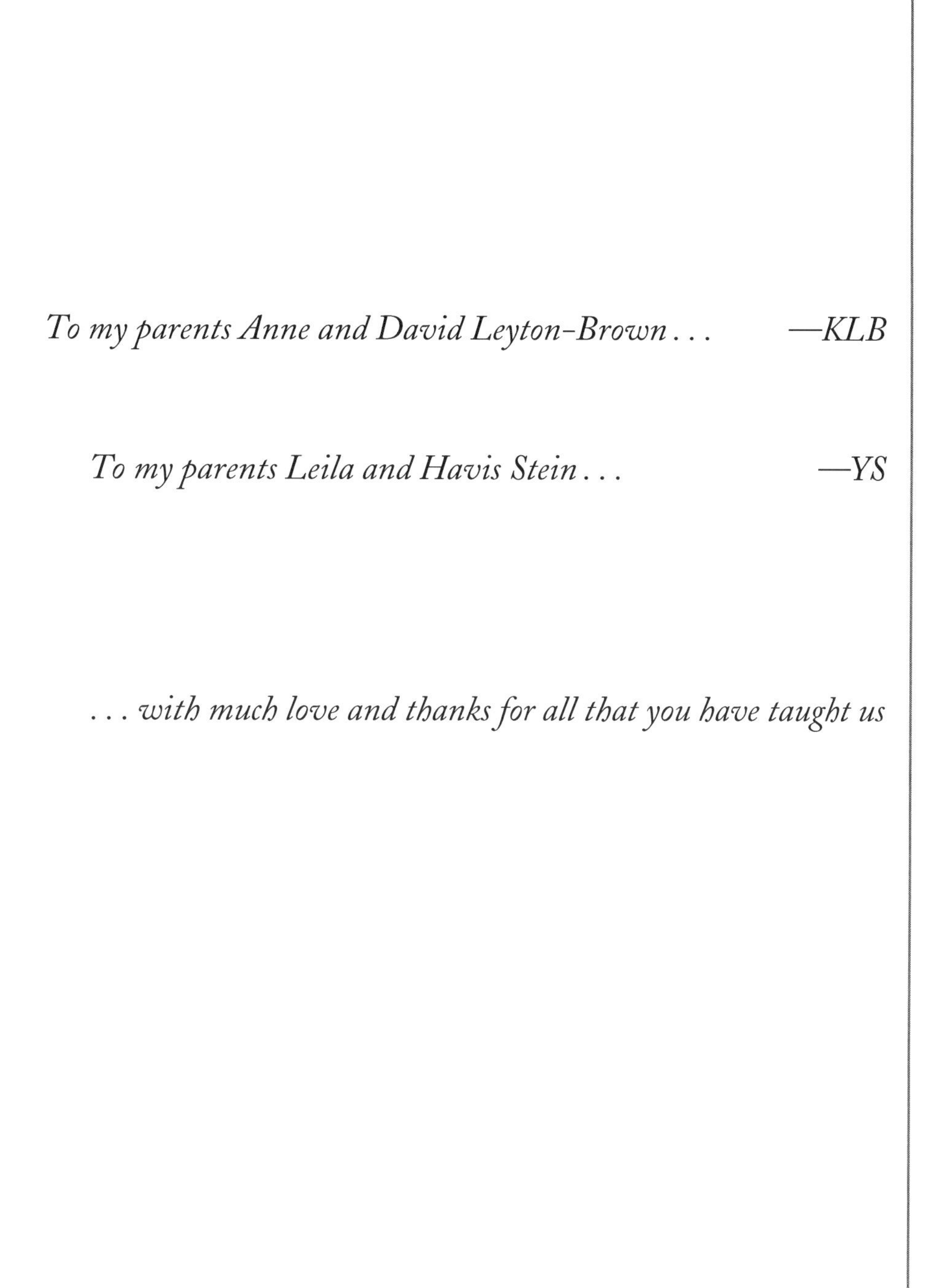

To my parents Anne and David Leyton-Brown . . . —*KLB*

To my parents Leila and Havis Stein . . . —*YS*

. . . with much love and thanks for all that you have taught us

Contents

Credits and Acknowledgments xiii

Preface xv

1. Games in Normal Form 1
   - 1.1 Example: The TCP User's Game 2
   - 1.2 Definition of Games in Normal Form 3
   - 1.3 More Examples of Normal-Form Games 4
     - 1.3.1 Prisoner's Dilemma 4
     - 1.3.2 Common-payoff Games 4
     - 1.3.3 Zero-sum Games 5
     - 1.3.4 Battle of the Sexes 7
   - 1.4 Strategies in Normal-form Games 7

2. Analyzing Games: From Optimality To Equilibrium 9
   - 2.1 Pareto optimality 9
   - 2.2 Defining Best Response and Nash Equilibrium 10
   - 2.3 Finding Nash Equilibria 11

3. Further Solution Concepts for Normal-Form Games 15
   - 3.1 Maxmin and Minmax Strategies 15
   - 3.2 Minimax Regret 18
   - 3.3 Removal of Dominated Strategies 20
   - 3.4 Rationalizability 23
   - 3.5 Correlated Equilibrium 24
   - 3.6 Trembling-Hand Perfect Equilibrium 26
   - 3.7 ϵ-Nash Equilibrium 27
   - 3.8 Evolutionarily Stable Strategies 28

4. Games With Sequential Actions: The Perfect-information Extensive Form 31
   - 4.1 Definition 31
   - 4.2 Strategies and Equilibria 32
   - 4.3 Subgame-Perfect Equilibrium 35
   - 4.4 Backward Induction 38

5. Generalizing the Extensive Form: Imperfect-Information Games 41
- 5.1 Definition 41
- 5.2 Strategies and Equilibria 42
- 5.3 Sequential Equilibrium 45

6. Repeated and Stochastic Games 49
- 6.1 Finitely Repeated Games 49
- 6.2 Infinitely Repeated Games 50
- 6.3 Stochastic Games 53
  - 6.3.1 Definition 53
  - 6.3.2 Strategies and Equilibria 54

7. Uncertainty About Payoffs: Bayesian Games 57
- 7.1 Definition 59
  - 7.1.1 Information Sets 59
  - 7.1.2 Extensive Form with Chance Moves 60
  - 7.1.3 Epistemic Types 61
- 7.2 Strategies and Equilibria 61
- 7.3 Computing Equilibria 64
- 7.4 *Ex-post* Equilibria 67

8. Coalitional Game Theory 69
- 8.1 Coalitional Games with Transferable Utility 69
- 8.2 Classes of Coalitional Games 70
- 8.3 Analyzing Coalitional Games 72
  - 8.3.1 The Shapley Value 73
  - 8.3.2 The Core 75

History and References 79

References 83

Index 85

Credits and Acknowledgments

We thank the many colleagues, including past and present graduate students, who made substantial contributions. Sam Ieong deserves special mention: Chapter 8 (coalitional games) is based entirely on writing by him, and he was also closely involved in the editing of this chapter. Other colleagues either supplied material or provided useful council. These include Felix Brandt, Vince Conitzer, Yossi Feinberg, Jeff Fletcher, Nando de Freitas, Ben Galin, Geoff Gordon, Teg Grenager, Albert Xin Jiang, David Poole, Peter Stone, David Thompson, Bob Wilson, and Erik Zawadzki. We also thank David Thompson for his assistance with the production of this book, particularly with the index and bibliography.

Of course, none of our colleagues are to be held accountable for any errors or other shortcomings. We claim sole credit for those.

We thank Morgan & Claypool, and particularly our editor Mike Morgan, for publishing *Essentials of Game Theory*, and indeed for suggesting the project in the first place. This booklet weaves together excerpts from our much longer book, *Multiagent Systems: Algorithmic, Game-Theoretic and Logical Foundations*, published by Cambridge University Press. We thank CUP, and particularly our editor Lauren Cowles, for not only agreeing to but indeed supporting the publication of this booklet. We are fortunate to be working with such stellar, forward-looking editors and publishers.

A great many additional colleagues contributed to the full Multiagent Systems book, and we thank them there. Since the project has been in the works in one way or another since 2001, it is possible—indeed, likely—that we have failed to thank some people. We apologize deeply in advance.

Last, and certainly not least, we thank our families, for supporting us through this time-consuming project. We dedicate this book to them, with love.

Preface

Game theory is the mathematical study of interaction among independent, self-interested agents. It is studied primarily by mathematicians and economists, microeconomics being its main initial application area. So what business do two computer scientists have publishing a text on game theory?

The origin of this booklet is our much longer book, *Multiagent Systems: Algorithmic, Game-Theoretic, and Logical Foundations*, which covers diverse theories relevant to the broad area of Multiagent Systems within Artificial Intelligence (AI) and other areas of computer science. Like many other disciplines, computer science—and in particular AI—have been profoundly influenced by game theory, with much back and forth between the fields taking place in recent years. And so it is not surprising to find that *Multiagent Systems* contains a fair bit of material on game theory. That material can be crudely divided into two kinds: basics, and more advanced material relevant to AI and computer science. This booklet weaves together the material of the first kind.

Many textbooks on game theory exist, some of them superb. The serious student of game theory cannot afford to neglect those, and in the final chapter we provide some references. But the audience for game theory has grown dramatically in recent years, spanning disciplines as diverse as political science, biology, psychology, linguistics, sociology—and indeed computer science—among many others. What has been missing is a relatively short introduction to the field covering the common basis that any one interested in game theory is likely to require. Such a text would minimize notation, ruthlessly focus on essentials, and yet not sacrifice rigor. This booklet aims to fill this gap. It is the book we wish *we* had had when we first ventured into the field.

We should clarify what we mean by "essentials." We cover the main classes of games, their representations, and the main concepts used to analyze them (so-called "solution concepts"). We cannot imagine any consumer of game theory who will not require a solid grounding in each of these topics. We discuss them in sufficient depth to provide this grounding, though of course much more can be said about each of them. This leaves out many topics in game theory that are key in certain applications, but not in all. Some examples are computational aspects of games and computationally motivated representations, learning in games, and mechanism design (in particular, auction theory). By omitting these topics we do not mean to suggest that they are unimportant, only that they will not be equally relevant to *everyone* who finds use for game theory. The reader of this booklet will likely be grounded in a particular discipline, and will thus need to augment his or her reading with material essential to that discipline.

This book makes an appropriate text for an advanced undergraduate course or a game theory unit in a graduate course. The book's Web site

http : //www.gtessentials.org

contains additional resources for both students and instructors.

A final word on pronouns and gender. We use male pronouns to refer to agents throughout the book. We debated this between us, not being happy with any of the alternatives. In the end we reluctantly settled on the "standard" male convention rather than the reverse female convention or the grammatically-dubious "they." We urge the reader not to read patriarchal intentions into our choice.

CHAPTER 1

Games in Normal Form

Game theory studies what happens when self-interested agents interact. What does it mean to say that agents are self-interested? It does not necessarily mean that they want to cause harm to each other, or even that they care only about themselves. Instead, it means that each agent has his own description of which states of the world he likes—which can include good things happening to other agents—and that he acts in an attempt to bring about these states of the world.

The dominant approach to modeling an agent's interests is *utility theory*. This theoretical approach quantifies an agent's degree of preference across a set of available alternatives, and describes how these preferences change when an agent faces uncertainty about which alternative he will receive. Specifically, a *utility function* is a mapping from states of the world to real numbers. These numbers are interpreted as measures of an agent's level of happiness in the given states. When the agent is uncertain about which state of the world he faces, his utility is defined as the expected value of his utility function with respect to the appropriate probability distribution over states.

When agents have utility functions, acting optimally in an uncertain environment is conceptually straightforward—at least as long as the outcomes and their probabilities are known to the agent and can be succinctly represented. However, things can get considerably more complicated when the world contains *two or more* utility-maximizing agents whose actions can affect each other's utilities. To study such settings, we must turn to *noncooperative* game theory.

The term "noncooperative" could be misleading, since it may suggest that the theory applies exclusively to situations in which the interests of different agents conflict. This is not the case, although it is fair to say that the theory is most interesting in such situations. By the same token, in Chapter 8 we will see that *coalitional game theory* (also known as *cooperative game theory*) does not apply only in situations in which agents' interests align with each other. The essential difference between the two branches is that in noncooperative game theory the basic modeling unit is the individual (including his beliefs, preferences, and possible actions) while in coalitional game theory the basic modeling unit is the group. We will return to that later in Chapter 8, but for now let us proceed with the individualistic approach.

	C	D
C	$-1, -1$	$-4, 0$
D	$0, -4$	$-3, -3$

FIGURE 1.1: The TCP user's (aka the Prisoner's) Dilemma.

1.1 EXAMPLE: THE TCP USER'S GAME

Let us begin with a simpler example to provide some intuition about the type of phenomena we would like to study. Imagine that you and another colleague are the only people using the internet. Internet traffic is governed by the TCP protocol. One feature of TCP is the *backoff* mechanism; if the rates at which you and your colleague send information packets into the network causes congestion, you each back off and reduce the rate for a while until the congestion subsides. This is how a correct implementation works. A defective one, however, will not back off when congestion occurs. You have two possible strategies: C (for using a correct implementation) and D (for using a defective one). If both you and your colleague adopt C then your average packet delay is 1 ms. If you both adopt D the delay is 3 ms, because of additional overhead at the network router. Finally, if one of you adopts D and the other adopts C then the D adopter will experience no delay at all, but the C adopter will experience a delay of 4 ms.

These consequences are shown in Figure 1.1. Your options are the two rows, and your colleague's options are the columns. In each cell, the first number represents your payoff (or, the negative of your delay) and the second number represents your colleague's payoff.[1]

Given these options what should you adopt, C or D? Does it depend on what you think your colleague will do? Furthermore, from the perspective of the network operator, what kind of behavior can he expect from the two users? Will any two users behave the same when presented with this scenario? Will the behavior change if the network operator allows the users to communicate with each other before making a decision? Under what changes to the delays would the users' decisions still be the same? How would the users behave if they have the opportunity to face this same decision with the same counterpart multiple times? Do answers to these questions depend on how rational the agents are and how they view each other's rationality?

[1] A more standard name for this game is the Prisoner's Dilemma; we return to this in Section 1.3.1.

Game theory gives answers to many of these questions. It tells us that any rational user, when presented with this scenario once, will adopt D—regardless of what the other user does. It tells us that allowing the users to communicate beforehand will not change the outcome. It tells us that for perfectly rational agents, the decision will remain the same even if they play multiple times; however, if the number of times that the agents will play this is infinite, or even uncertain, we may see them adopt C.

1.2 DEFINITION OF GAMES IN NORMAL FORM

The normal form, also known as the strategic or matrix form, is the most familiar representation of strategic interactions in game theory. A game written in this way amounts to a representation of every player's utility for every state of the world, in the special case where states of the world depend only on the players' combined actions. Consideration of this special case may seem uninteresting. However, it turns out that settings in which the state of the world also depends on randomness in the environment—called Bayesian games and introduced in Chapter 7—can be reduced to (much larger) normal-form games. Indeed, there also exist normal-form reductions for other game representations, such as games that involve an element of time (extensive-form games, introduced in Chapter 4). Because most other representations of interest can be reduced to it, the normal-form representation is arguably the most fundamental in game theory.

Definition 1.2.1 (Normal-form game). *A (finite,* n*-person)* normal-form game *is a tuple* (N, A, u), *where:*

- N *is a finite set of* n *players, indexed by* i;
- $A = A_1 \times \cdots \times A_n$, *where* A_i *is a finite set of* actions *available to player* i. *Each vector* $a = (a_1, \ldots, a_n) \in A$ *is called an* action profile;
- $u = (u_1, \ldots, u_n)$ *where* $u_i : A \mapsto \mathbb{R}$ *is a real-valued* utility *(or* payoff*)* function *for player* i.

A natural way to represent games is via an n-dimensional matrix. We already saw a two-dimensional example in Figure 1.1. In general, each row corresponds to a possible action for player 1, each column corresponds to a possible action for player 2, and each cell corresponds to one possible outcome. Each player's utility for an outcome is written in the cell corresponding to that outcome, with player 1's utility listed first.

1.3 MORE EXAMPLES OF NORMAL-FORM GAMES

1.3.1 Prisoner's Dilemma

Previously, we saw an example of a game in normal form, namely, the Prisoner's (or the TCP user's) Dilemma. However, it turns out that the precise payoff numbers play a limited role. The

	C	D
C	a, a	b, c
D	c, b	d, d

FIGURE 1.2: Any $c > a > d > b$ define an instance of Prisoner's Dilemma.

essence of the Prisoner's Dilemma example would not change if the -4 was replaced by -5, or if 100 was added to each of the numbers.[2] In its most general form, the Prisoner's Dilemma is any normal-form game shown in Figure 1.2, in which $c > a > d > b$.[3]

Incidentally, the name "Prisoner's Dilemma" for this famous game-theoretic situation derives from the original story accompanying the numbers. The players of the game are two prisoners suspected of a crime rather than two network users. The prisoners are taken to separate interrogation rooms, and each can either "confess" to the crime or "deny" it (or, alternatively, "cooperate" or "defect"). If the payoff are all nonpositive, their absolute values can be interpreted as the length of jail term each of prisoner will get in each scenario.

1.3.2 Common-payoff Games

There are some restricted classes of normal-form games that deserve special mention. The first is the class of *common-payoff games*. These are games in which, for every action profile, all players have the same payoff.

Definition 1.3.1 (Common-payoff game). *A* common-payoff game *is a game in which for all action profiles $a \in A_1 \times \cdots \times A_n$ and any pair of agents i, j, it is the case that $u_i(a) = u_j(a)$.*

Common-payoff games are also called *pure coordination games* or *team games*. In such games the agents have no conflicting interests; their sole challenge is to coordinate on an action that is maximally beneficial to all.

As an example, imagine two drivers driving towards each other in a country having no traffic rules, and who must independently decide whether to drive on the left or on the right. If the drivers choose the same side (left or right) they have some high utility, and otherwise they have a low utility. The game matrix is shown in Figure 1.3.

[2]More generally, under standard utility theory games are are insensitive to any positive affine transformation of the payoffs. This means that one can replace each payoff x by $ax + b$, for any fixed real numbers $a > 0$ and b.
[3]Under some definitions, there is the further requirement that $a > \frac{b+c}{2}$, which guarantees that the outcome (C, C) maximizes the sum of the agents' utilities.

	Left	Right
Left	$1,1$	$0,0$
Right	$0,0$	$1,1$

FIGURE 1.3: Coordination game.

1.3.3 Zero-sum Games

At the other end of the spectrum from pure coordination games lie *zero-sum games*, which (bearing in mind the comment we made earlier about positive affine transformations) are more properly called *constant-sum games.* Unlike common-payoff games, constant-sum games are meaningful primarily in the context of two-player (though not necessarily two-strategy) games.

Definition 1.3.2 (Constant-sum game). *A two-player normal-form game is* constant-sum *if there exists a constant c such that for each strategy profile $a \in A_1 \times A_2$ it is the case that $u_1(a) + u_2(a) = c$.*

For convenience, when we talk of constant-sum games going forward we will always assume that $c = 0$, that is, that we have a zero-sum game. If common-payoff games represent situations of pure coordination, zero-sum games represent situations of pure competition; one player's gain must come at the expense of the other player. The reason zero-sum games are most meaningful for two agents is that if you allow more agents, any game can be turned into a zero-sum game by adding a dummy player whose actions do not impact the payoffs to the other agents, and whose own payoffs are chosen to make the sum of payoffs in each outcome zero.

A classical example of a zero-sum game is the game of *Matching Pennies.* In this game, each of the two players has a penny, and independently chooses to display either heads or tails. The two players then compare their pennies. If they are the same then player 1 pockets both, and otherwise player 2 pockets them. The payoff matrix is shown in Figure 1.4.

The popular children's game of Rock, Paper, Scissors, also known as Rochambeau, provides a three-strategy generalization of the matching-pennies game. The payoff matrix of this zero-sum game is shown in Figure 1.5. In this game, each of the two players can choose either rock, paper, or scissors. If both players choose the same action, there is no winner and the utilities are zero. Otherwise, each of the actions wins over one of the other actions and loses to the other remaining action.

	Heads	Tails
Heads	$1, -1$	$-1, 1$
Tails	$-1, 1$	$1, -1$

FIGURE 1.4: Matching Pennies game.

1.3.4 Battle of the Sexes

In general, games tend to include elements of both coordination and competition. Prisoner's Dilemma does, although in a rather paradoxical way. Here is another well-known game that includes both elements. In this game, called *Battle of the Sexes*, a husband and wife wish to go to the movies, and they can select among two movies: "Lethal Weapon (LW)" and "Wondrous Love (WL)." They much prefer to go together rather than to separate movies, but while the wife (player 1) prefers LW, the husband (player 2) prefers WL. The payoff matrix is shown in Figure 1.6. We will return to this game shortly.

1.4 STRATEGIES IN NORMAL-FORM GAMES

We have so far defined the actions available to each player in a game, but not yet his set of *strategies* or his available choices. Certainly one kind of strategy is to select a single action and play it. We call such a strategy a *pure strategy*, and we will use the notation we have already developed for actions to represent it. We call a choice of pure strategy for each agent a *pure-strategy profile*.

	Rock	Paper	Scissors
Rock	$0, 0$	$-1, 1$	$1, -1$
Paper	$1, -1$	$0, 0$	$-1, 1$
Scissors	$-1, 1$	$1, -1$	$0, 0$

FIGURE 1.5: Rock, Paper, Scissors game.

		Husband	
		LW	WL
Wife	LW	$2, 1$	$0, 0$
	WL	$0, 0$	$1, 2$

FIGURE 1.6: Battle of the Sexes game.

Players could also follow another, less obvious type of strategy: randomizing over the set of available actions according to some probability distribution. Such a strategy is called a mixed strategy. Although it may not be immediately obvious why a player should introduce randomness into his choice of action, in fact in a multiagent setting the role of mixed strategies is critical.

We define a mixed strategy for a normal-form game as follows.

Definition 1.4.1 (Mixed strategy). *Let (N, A, u) be a normal-form game, and for any set X let $\Pi(X)$ be the set of all probability distributions over X. Then the set of* mixed strategies *for player i is $S_i = \Pi(A_i)$.*

Definition 1.4.2 (Mixed-strategy profile). *The set of* mixed-strategy profiles *is simply the Cartesian product of the individual mixed-strategy sets, $S_1 \times \cdots \times S_n$.*

By $s_i(a_i)$ we denote the probability that an action a_i will be played under mixed strategy s_i. The subset of actions that are assigned positive probability by the mixed strategy s_i is called the *support* of s_i.

Definition 1.4.3 (Support). *The* support *of a mixed strategy s_i for a player i is the set of pure strategies $\{a_i | s_i(a_i) > 0\}$.*

Note that a pure strategy is a special case of a mixed strategy, in which the support is a single action. At the other end of the spectrum we have *fully mixed strategies*. A strategy is fully mixed if it has full support (i.e., if it assigns every action a nonzero probability).

We have not yet defined the payoffs of players given a particular strategy profile, since the payoff matrix defines those directly only for the special case of pure-strategy profiles. But the generalization to mixed strategies is straightforward, and relies on the basic notion of decision theory—*expected utility*. Intuitively, we first calculate the probability of reaching each outcome given the strategy profile, and then we calculate the average of the payoffs of the outcomes, weighted by the probabilities of each outcome. Formally, we define the expected utility as follows (overloading notation, we use u_i for both utility and expected utility).

Definition 1.4.4 (Expected utility of a mixed strategy). *Given a normal-form game (N, A, u), the expected utility u_i for player i of the mixed-strategy profile $s = (s_1, \ldots, s_n)$ is defined as*

$$u_i(s) = \sum_{a \in A} u_i(a) \prod_{j=1}^{n} s_j(a_j).$$

CHAPTER 2

Analyzing Games: From Optimality To Equilibrium

Now that we have defined what games in normal form are and what strategies are available to players in them, the question is how to reason about such games. In single-agent decision theory the key notion is that of an *optimal strategy*, that is, a strategy that maximizes the agent's expected payoff for a given environment in which the agent operates. The situation in the single-agent case can be fraught with uncertainty, since the environment might be stochastic, partially observable, and spring all kinds of surprises on the agent. However, the situation is even more complex in a multiagent setting. In this case the environment includes—or, in many cases we discuss, consists entirely of—other agents, all of whom are also hoping to maximize their payoffs. Thus the notion of an optimal strategy for a given agent is not meaningful; the best strategy depends on the choices of others.

Game theorists deal with this problem by identifying certain subsets of outcomes, called *solution concepts*, that are interesting in one sense or another. In this section we describe two of the most fundamental solution concepts: Pareto optimality and Nash equilibrium.

2.1 PARETO OPTIMALITY

First, let us investigate the extent to which a notion of optimality can be meaningful in games. From the point of view of an outside observer, can some outcomes of a game be said to be better than others?

This question is complicated because we have no way of saying that one agent's interests are more important than another's. For example, it might be tempting to say that we should prefer outcomes in which the sum of agents' utilities is higher. However, as remarked in Footnote 2 earlier, we can apply any positive affine transformation to an agent's utility function and obtain another valid utility function. For example, we could multiply all of player 1's payoffs by 1,000—this could clearly change which outcome maximized the sum of agents' utilities.

Thus, our problem is to find a way of saying that some outcomes are better than others, even when we only know agents' utility functions up to a positive affine transformation. Imagine that each agent's utility is a monetary payment that you will receive, but that each payment comes in a different currency, and you do not know anything about the exchange rates. Which outcomes should you prefer? Observe that, while it is not usually possible to identify the best outcome, there *are* situations in which you can be sure that one outcome is better than another. For example, it is better to get 10 units of currency A and 3 units of currency B than to get 9 units of currency A and 3 units of currency B, regardless of the exchange rate. We formalize this intuition in the following definition.

Definition 2.1.1 (Pareto domination). *Strategy profile s* Pareto dominates *strategy profile s' if for all $i \in N$, $u_i(s) \geq u_i(s')$, and there exists some $j \in N$ for which $u_j(s) > u_j(s')$.*

In other words, in a Pareto-dominated strategy profile some player can be made better off without making any other player worse off. Observe that we define Pareto domination over strategy profiles, not just action profiles.

Pareto domination gives us a partial ordering over strategy profiles. Thus, in answer to our question before, we cannot generally identify a single "best" outcome; instead, we may have a set of noncomparable optima.

Definition 2.1.2 (Pareto optimality). *Strategy profile s is* Pareto optimal, *or* strictly Pareto efficient, *if there does not exist another strategy profile $s' \in S$ that Pareto dominates s.*

We can easily draw several conclusions about Pareto optimal strategy profiles. First, every game must have at least one such optimum, and there must always exist at least one such optimum in which all players adopt pure strategies. Second, some games will have multiple optima. For example, in zero-sum games, *all* strategy profiles are strictly Pareto efficient. Finally, in common-payoff games, all Pareto optimal strategy profiles have the same payoffs.

2.2 DEFINING BEST RESPONSE AND NASH EQUILIBRIUM

Now we will look at games from an individual agent's point of view, rather than from the vantage point of an outside observer. This will lead us to the most influential solution concept in game theory, the *Nash equilibrium*.

Our first observation is that if an agent knew how the others were going to play, his strategic problem would become simple. Specifically, he would be left with the single-agent problem of choosing a utility-maximizing action. Formally, define $s_{-i} = (s_1, \ldots, s_{i-1}, s_{i+1}, \ldots, s_n)$, a strategy profile s without agent i's strategy. Thus we can write $s = (s_i, s_{-i})$. If the agents other than i (whom we denote $-i$) were to commit to play s_{-i}, a utility-maximizing agent i would face the problem of determining his best response.

Definition 2.2.1 (Best response). *Player i's* best response *to the strategy profile s_{-i} is a mixed strategy $s_i^* \in S_i$ such that $u_i(s_i^*, s_{-i}) \geq u_i(s_i, s_{-i})$ for all strategies $s_i \in S_i$.*

The best response is not necessarily unique. Indeed, except in the extreme case in which there is a unique best response that is a pure strategy, the number of best responses is always infinite. When the support of a best response s^* includes two or more actions, the agent must be indifferent among them—otherwise, the agent would prefer to reduce the probability of playing at least one of the actions to zero. But thus *any* mixture of these actions must also be a best response, not only the particular mixture in s^*. Similarly, if there are two pure strategies that are individually best responses, any mixture of the two is necessarily also a best response.

Of course, in general an agent will not know what strategies the other players will adopt. Thus, the notion of best response is not a solution concept—it does not identify an interesting set of outcomes in this general case. However, we can leverage the idea of best response to define what is arguably the most central notion in noncooperative game theory, the Nash equilibrium.

Definition 2.2.2 (Nash equilibrium). *A strategy profile $s = (s_1, \ldots, s_n)$ is a* Nash equilibrium *if, for all agents i, s_i is a best response to s_{-i}.*

Intuitively, a Nash equilibrium is a *stable* strategy profile: no agent would want to change his strategy if he knew what strategies the other agents were following.

We can divide Nash equilibria into two categories, strict and weak, depending on whether or not every agent's strategy constitutes a *unique* best response to the other agents' strategies.

Definition 2.2.3 (Strict Nash). *A strategy profile $s = (s_1, \ldots, s_n)$ is a if, for all agents i and for all strategies $s_i' \neq s_i$, $u_i(s_i, s_{-i}) > u_i(s_i', s_{-i})$.*

Definition 2.2.4 (Weak Nash). *A strategy profile $s = (s_1, \ldots, s_n)$ is a if, for all agents i and for all strategies $s_i' \neq s_i$, $u_i(s_i, s_{-i}) \geq u_i(s_i', s_{-i})$, and s is not a strict Nash equilibrium.*

Intuitively, weak Nash equilibria are less stable than strict Nash equilibria, because in the former case at least one player has a best response to the other players' strategies that is not his equilibrium strategy. Mixed-strategy Nash equilibria are necessarily always weak, while pure-strategy Nash equilibria can be either strict or weak, depending on the game.

2.3 FINDING NASH EQUILIBRIA

Consider again the Battle of the Sexes game. We immediately see that it has two pure-strategy Nash equilibria, depicted in Figure 2.1.

We can check that these are Nash equilibria by confirming that whenever one of the players plays the given (pure) strategy, the other player would only lose by deviating.

	LW	WL
LW	2, 1	0, 0
WL	0, 0	1, 2

FIGURE 2.1: Pure-strategy Nash equilibria in the Battle of the Sexes game.

Are these the only Nash equilibria? The answer is no; although they are indeed the only pure-strategy equilibria, there is also another mixed-strategy equilibrium. In general, it is tricky to compute a game's mixed-strategy equilibria. This is a weighty topic lying outside the scope of this booklet (but see, for example, Chapter 4 of Shoham and Leyton-Brown [2008]). However, we will show here that this computational problem is easy when we know (or can guess) the *support* of the equilibrium strategies, particularly so in this small game. Let us now guess that both players randomize, and let us assume that husband's strategy is to play LW with probability p and WL with probability $1 - p$. Then if the wife, the row player, also mixes between her two actions, she must be indifferent between them, given the husband's strategy. (Otherwise, she would be better off switching to a pure strategy according to which she only played the better of her actions.) Then we can write the following equations.

$$\begin{aligned} U_{\text{wife}}(\text{LW}) &= U_{\text{wife}}(\text{WL}) \\ 2 * p + 0 * (1 - p) &= 0 * p + 1 * (1 - p) \\ p &= \frac{1}{3} \end{aligned}$$

We get the result that in order to make the wife indifferent between her actions, the husband must choose LW with probability 1/3 and WL with probability 2/3. Of course, since the husband plays a mixed strategy he must also be indifferent between his actions. By a similar calculation it can be shown that to make the husband indifferent, the wife must choose LW with probability 2/3 and WL with probability 1/3. Now we can confirm that we have indeed found an equilibrium: since both players play in a way that makes the other indifferent between their actions, they are both best responding to each other. Like all mixed-strategy equilibria, this is a weak Nash equilibrium. The expected payoff of both agents is 2/3 in this equilibrium, which means that each of the pure-strategy equilibria Pareto-dominates the mixed-strategy equilibrium.

	Heads	Tails
Heads	$1, -1$	$-1, 1$
Tails	$-1, 1$	$1, -1$

FIGURE 2.2: The Matching Pennies game.

Earlier, we mentioned briefly that mixed strategies play an important role. The previous example may not make it obvious, but now consider again the Matching Pennies game, reproduced in Figure 2.2. It is not hard to see that no pure strategy could be part of an equilibrium in this game of pure competition. Therefore, likewise there can be no strict Nash equilibrium in this game. But using the aforementioned procedure, the reader can verify that again there exists a mixed-strategy equilibrium; in this case, each player chooses one of the two available actions with probability 1/2.

We have now seen two examples in which we managed to find Nash equilibria (three equilibria for Battle of the Sexes, one equilibrium for Matching Pennies). Did we just luck out? Here there is some good news—it was not just luck.

Theorem 2.3.1 (Nash, 1951). *Every game with a finite number of players and action profiles has at least one Nash equilibrium.*

The proof of this result is somewhat involved, and we do not discuss it here except to mention that it is typically achieved by appealing to a *fixed-point theorem* from mathematics, such as those due to Kakutani and Brouwer (a detailed proof appears, for example, in Chapter 3 of Shoham and Leyton-Brown [2008]).

Nash's theorem depends critically on the availability of mixed strategies to the agents. (Many games, such as Matching Pennies, have only mixed-strategy equilibria.) However, what does it mean to say that an agent plays a mixed-strategy Nash equilibrium? Do players really sample probability distributions in their heads? Some people have argued that they really do. One well-known motivating example for mixed strategies involves soccer: specifically, a kicker and a goalie getting ready for a penalty kick. The kicker can kick to the left or the right, and the goalie can jump to the left or the right. The kicker scores if and only if he kicks to one side and the goalie jumps to the other; this is thus best modeled as Matching Pennies. Any pure strategy on the part of either player invites a winning best response on the part of the other player. It is only by kicking or jumping in either direction with equal probability, goes the argument, that the opponent cannot exploit your strategy.

Of course, this argument is not uncontroversial. In particular, it can be argued that the strategies of each player are deterministic, but each player has uncertainty regarding the other player's strategy. This is indeed a second possible interpretation of mixed strategies: the mixed strategy of player i is everyone else's assessment of how likely i is to play each pure strategy. In equilibrium, i's mixed strategy has the further property that every action in its support is a best response to player i's beliefs about the other agents' strategies.

Finally, there are two interpretations that are related to learning in multiagent systems. In one interpretation, the game is actually played many times repeatedly, and the probability of a pure strategy is the fraction of the time it is played in the limit (its so-called *empirical frequency*). In the other interpretation, not only is the game played repeatedly, but each time it involves two different agents selected at random from a large population. In this interpretation, each agent in the population plays a pure strategy, and the probability of a pure strategy represents the fraction of agents playing that strategy.

CHAPTER 3

Further Solution Concepts for Normal-Form Games

As described earlier at the beginning of Chapter 2, we reason about multiplayer games using *solution concepts*, principles according to which we identify interesting subsets of the outcomes of a game. While the most important solution concept is the Nash equilibrium, there are also a large number of others, only some of which we will discuss here. Some of these concepts are more restrictive than the Nash equilibrium, some less so, and some noncomparable. In subsequent chapters we will introduce some additional solution concepts that are only applicable to game representations other than the normal form.

3.1 MAXMIN AND MINMAX STRATEGIES

The *maxmin strategy* of player i in an n-player, general-sum game is a (not necessarily unique, and in general mixed) strategy that maximizes i's worst-case payoff, in the situation where all the other players happen to play the strategies which cause the greatest harm to i. The *maxmin value* (or *security level*) of the game for player i is that minimum amount of payoff guaranteed by a maxmin strategy.

Definition 3.1.1 (Maxmin). *The* maxmin strategy *for player i is* $\arg\max_{s_i} \min_{s_{-i}} u_i(s_i, s_{-i})$, *and the* maxmin value *for player i is* $\max_{s_i} \min_{s_{-i}} u_i(s_i, s_{-i})$.

Although the maxmin strategy is a concept that makes sense in simultaneous-move games, it can be understood through the following temporal intuition. The maxmin strategy is i's best choice when first i must commit to a (possibly mixed) strategy, and then the remaining agents $-i$ observe this strategy (but not i's action choice) and choose their own strategies to minimize i's expected payoff. In the Battle of the Sexes game (Figure 1.6), the maxmin value for either player is 2/3, and requires the maximizing agent to play a mixed strategy. (Do you see why?)

While it may not seem reasonable to assume that the other agents would be solely interested in minimizing i's utility, it is the case that i plays a maxmin strategy and the other agents play arbitrarily, i will still receive an expected payoff of at least his maxmin value. This means that the maxmin strategy is a sensible choice for a conservative agent who wants to maximize his expected utility without having to make any assumptions about the other agents, such as that they will act rationally according to their own interests, or that they will draw their action choices from some known distributions.

The *minmax strategy* and *minmax value* play a dual role to their maxmin counterparts. In two-player games the minmax strategy for player i against player $-i$ is a strategy that keeps the maximum payoff of $-i$ at a minimum, and the minmax value of player $-i$ is that minimum. This is useful when we want to consider the amount that one player can punish another without regard for his own payoff. Such punishment can arise in repeated games, as we will see in Section 6. The formal definitions follow.

Definition 3.1.2 (Minmax, two-player). *In a two-player game, the* minmax strategy *for player i against player $-i$ is* $\arg\min_{s_i} \max_{s_{-i}} u_{-i}(s_i, s_{-i})$, *and player $-i$'s* minmax value *is* $\min_{s_i} \max_{s_{-i}} u_{-i}(s_i, s_{-i})$.

In n-player games with $n > 2$, defining player i's minmax strategy against player j is a bit more complicated. This is because i will not usually be able to guarantee that j achieves minimal payoff by acting unilaterally. However, if we assume that all the players other than j choose to "gang up" on j—and that they are able to coordinate appropriately when there is more than one strategy profile that would yield the same minimal payoff for j—then we can define minmax strategies for the n-player case.

Definition 3.1.3 (Minmax, n-player). *In an n-player game, the* minmax strategy *for player i against player $j \neq i$ is i's component of the mixed-strategy profile s_{-j} in the expression* $\arg\min_{s_{-j}} \max_{s_j} u_j(s_j, s_{-j})$, *where $-j$ denotes the set of players other than j. As before, the* minmax value *for player j is* $\min_{s_{-j}} \max_{s_j} u_j(s_j, s_{-j})$.

As before, we can give intuition for the minmax value through a temporal setting. Imagine that the agents $-i$ must commit to a (possibly mixed) strategy profile, to which i can then play a best response. Player i receives his minmax value if players $-i$ choose their strategies in order to minimize i's expected utility after he plays his best response.

In two-player games, a player's minmax value is always equal to his maxmin value. For games with more than two players a weaker condition holds: a player's maxmin value is always less than or equal to his minmax value. (Can you explain why this is?)

Since neither an agent's maxmin strategy nor his minmax strategy depend on the strategies that the other agents actually choose, the maxmin and minmax strategies give rise to solution concepts in a straightforward way. We will call a mixed-strategy profile $s = (s_1, s_2, \ldots)$ a *maxmin strategy profile* of a given game if s_1 is a maxmin strategy for player 1, s_2 is a maxmin strategy for player 2 and so on. In two-player games, we can also define *minmax strategy profiles* analogously. In two-player, zero-sum games, there is a very tight connection between minmax and maxmin strategy profiles. Furthermore, these solution concepts are also linked to the Nash equilibrium.

Theorem 3.1.4 (Minimax theorem (von Neumann, 1928)). *In any finite, two-player, zero-sum game, in any Nash equilibrium*[1] *each player receives a payoff that is equal to both his maxmin value and his minmax value.*

Why is the minmax theorem important? It demonstrates that maxmin strategies, minmax strategies and Nash equilibria coincide in two-player, zero-sum games. In particular, Theorem 3.1.4 allows us to conclude that in two-player, zero-sum games:

1. Each player's maxmin value is equal to his minmax value. By convention, the maxmin value for player 1 is called the *value of the game*;
2. For both players, the set of maxmin strategies coincides with the set of minmax strategies; and
3. Any maxmin strategy profile (or, equivalently, minmax strategy profile) is a Nash equilibrium. Furthermore, these are all the Nash equilibria. Consequently, all Nash equilibria have the same payoff vector (namely, those in which player 1 gets the value of the game).

For example, in the Matching Pennies game in Figure 1.4, the value of the game is 0. The unique Nash equilibrium consists of both players randomizing between heads and tails with equal probability, which is both the maxmin strategy and the minmax strategy for each player.

Nash equilibria in zero-sum games can be viewed graphically as a "saddle" in a high-dimensional space. At a saddle point, any deviation of the agent lowers his utility and increases the utility of the other agent. It is easy to visualize in the simple case in which each agent has two pure strategies. In this case the space of strategy profiles can be viewed as all points on the square between (0,0) and (1,1), with each point in the square describing the mixed strategy of each agent. The payoff to player 1 (and thus the negative of the payoff to player 2) is indeed a saddle-shaped, three-dimensional surface above this square. Figure 3.1 (left) gives a pictorial example, illustrating player 1's expected utility in Matching Pennies as a function of both players' probabilities of playing heads. Figure 3.1 (right) adds a plane at $z = 0$ to make it

[1]The attentive reader might wonder how a theorem from 1928 can use the term "Nash equilibrium," when Nash's work was published in 1950. Von Neumann used different terminology and proved the theorem in a different way; however, the given presentation is probably clearer in the context of modern game theory.

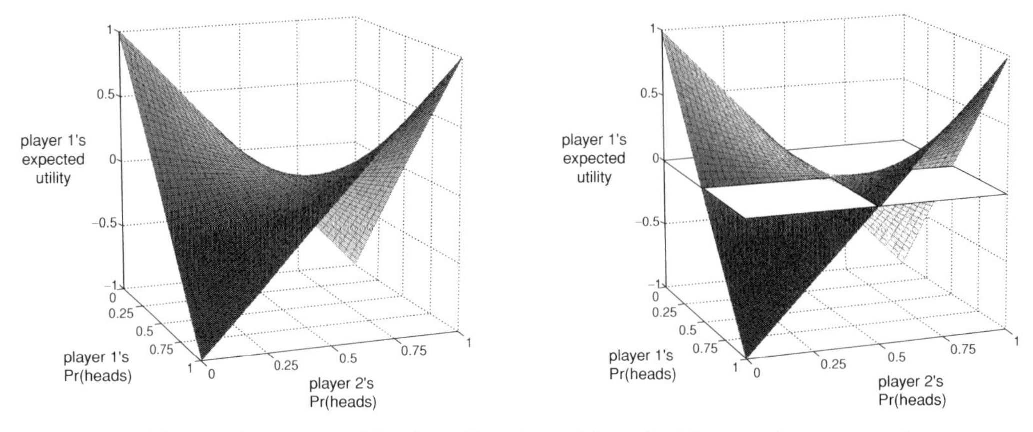

FIGURE 3.1: The saddle point in Matching Pennies, with and without a plane at $z = 0$.

easier to see that it is an equilibrium for both players to play heads 50% of the time and that zero is both the maxmin value and the minmax value for both players.

3.2 MINIMAX REGRET

We argued earlier that agents might play maxmin strategies in order to achieve good payoffs in the worst case, even in a game that is not zero sum. However, consider a setting in which the other agent is not believed to be malicious, but is instead believed to be entirely unpredictable. (Crucially, in this section we do not approach the problem as Bayesians, saying that agent i's beliefs can be described by a probability distribution; instead, we use a "pre-Bayesian" model in which i does not know such a distribution and indeed has no beliefs about it.) In such a setting, it can also make sense for agents to care about minimizing their worst-case *loss*, rather than simply maximizing their worst-case payoff.

Consider the game in Figure 3.2. Interpret the payoff numbers as pertaining to agent 1 only and let ϵ be an arbitrarily small positive constant. For this example it does not matter what agent 2's payoffs a, b, c, and d are, and we can even imagine that agent 1 does not know these values. Indeed, this could be one reason why player 1 would be unable to form beliefs about how player 2 would play, even if he were to believe that player 2 was rational. Let us imagine that agent 1 wants to determine a strategy to follow that makes sense despite his uncertainty about player 2. First, agent 1 might play his maxmin, or "safety level" strategy. In this game it is easy to see that player 1's maxmin strategy is to play B; this is because player 2's minmax strategy is to play R, and B is a best response to R.

If player 1 does not believe that player 2 is malicious, however, he might instead reason as follows. If player 2 were to play R then it would not matter very much how player 1 plays: the most he could lose by playing the wrong way would be ϵ. On the other hand, if player 2

	L	R
T	$100, a$	$1-\epsilon, b$
B	$2, c$	$1, d$

FIGURE 3.2: A game for contrasting maxmin with minimax regret. The numbers refer only to player 1's payoffs; ϵ is an arbitrarily small positive constant. Player 2's payoffs are the arbitrary (and possibly unknown) constants a, b, c, and d.

were to play L then player 1's action would be very significant: if player 1 were to make the wrong choice here then his utility would be decreased by 98. Thus player 1 might choose to play T in order to minimize his worst-case loss. Observe that this is the opposite of what he would choose if he followed his maxmin strategy.

Let us now formalize this idea. We begin with the notion of regret.

Definition 3.2.1 (Regret). *An agent i's* regret *for playing an action a_i if the other agents adopt action profile a_{-i} is defined as*

$$\left[\max_{a'_i \in A_i} u_i(a'_i, a_{-i})\right] - u_i(a_i, a_{-i}).$$

In words, this is the amount that i loses by playing a_i, rather than playing his best response to a_{-i}. Of course, i does not know what actions the other players will take; however, he can consider those actions that would give him the highest regret for playing a_i.

Definition 3.2.2 (Max regret). *An agent i's* maximum regret *for playing an action a_i is defined as*

$$\max_{a_{-i} \in A_{-i}} \left(\left[\max_{a'_i \in A_i} u_i(a'_i, a_{-i})\right] - u_i(a_i, a_{-i}) \right).$$

This is the amount that i loses by playing a_i rather than playing his best response to a_{-i}, if the other agents chose the a_{-i} that makes this loss as large as possible. Finally, i can choose his action in order to minimize this worst-case regret.

Definition 3.2.3 (Minimax regret). *Minimax regret actions for agent i are defined as*

$$\arg\min_{a_i \in A_i} \left[\max_{a_{-i} \in A_{-i}} \left(\left[\max_{a'_i \in A_i} u_i(a'_i, a_{-i})\right] - u_i(a_i, a_{-i}) \right) \right].$$

Thus, an agent's minimax regret action is an action that yields the smallest maximum regret. Minimax regret can be extended to a solution concept in the natural way, by identifying action profiles that consist of minimax regret actions for each player. Note that we can safely restrict ourselves to actions rather than mixed strategies in the definitions above (i.e., maximizing over the sets A_i and A_{-i} instead of S_i and S_{-i}), because of the linearity of expectation. We leave the proof of this fact as an exercise.

3.3 REMOVAL OF DOMINATED STRATEGIES

We first define what it means for one strategy to dominate another. Intuitively, one strategy dominates another for a player i if the first strategy yields i a greater payoff than the second strategy, for *any* strategy profile of the remaining players.[2] There are, however, three gradations of dominance, which are captured in the following definition.

Definition 3.3.1 (Domination). *Let s_i and s'_i be two strategies of player i, and S_{-i} the set of all strategy profiles of the remaining players. Then*

1. s_i strictly dominates s'_i *if for all $s_{-i} \in S_{-i}$, it is the case that $u_i(s_i, s_{-i}) > u_i(s'_i, s_{-i})$.*
2. s_i weakly dominates s'_i *if for all $s_{-i} \in S_{-i}$, it is the case that $u_i(s_i, s_{-i}) \geq u_i(s'_i, s_{-i})$, and for at least one $s_{-i} \in S_{-i}$, it is the case that $u_i(s_i, s_{-i}) > u_i(s'_i, s_{-i})$.*
3. s_i very weakly dominates s'_i *if for all $s_{-i} \in S_{-i}$, it is the case that $u_i(s_i, s_{-i}) \geq u_i(s'_i, s_{-i})$.*

If one strategy dominates all others, we say that it is (strongly, weakly or very weakly) *dominant.*

Definition 3.3.2 (Dominant strategy). *A strategy is* strictly (resp., weakly; very weakly) dominant *for an agent if it strictly (weakly; very weakly) dominates any other strategy for that agent.*

It is obvious that a strategy profile $(s_1, \ldots, s_n)$ in which every s_i is dominant for player i (whether strictly, weakly, or very weakly) is a Nash equilibrium. Such a strategy profile forms what is called an *equilibrium in dominant strategies* with the appropriate modifier (*strictly*, etc). An equilibrium in strictly dominant strategies is necessarily the unique Nash equilibrium. For example, consider again the Prisoner's Dilemma game. For each player, the strategy D is strictly dominant, and indeed (D, D) is the unique Nash equilibrium. Indeed, we can now explain the "dilemma" which is particularly troubling about the Prisoner's Dilemma game: the outcome reached in the unique equilibrium, which is an equilibrium in strictly dominant strategies, is also the only outcome that is *not* Pareto optimal.

[2]Note that here we consider strategy domination from one individual player's point of view; thus, this notion is unrelated to the concept of Pareto domination discussed earlier.

	L	C	R
U	$3,1$	$0,1$	$0,0$
M	$1,1$	$1,1$	$5,0$
D	$0,1$	$4,1$	$0,0$

FIGURE 3.3: A game with dominated strategies.

Games with dominant strategies play an important role in game theory, especially in games handcrafted by experts. This is true in particular in *mechanism design*, an area of game theory not covered in this booklet. However, dominant strategies are rare in naturally occurring games. More common are dominated strategies.

Definition 3.3.3 (Dominated strategy). *A strategy s_i is* strictly (weakly; very weakly) dominated *for an agent i if some other strategy s_i' strictly (weakly; very weakly) dominates s_i.*

Let us focus for the moment on strictly dominated strategies. Intuitively, all strictly dominated pure strategies can be ignored, since they can never be best responses to any moves by the other players. There are several subtleties, however. First, once a pure strategy is eliminated, another strategy that was not dominated can become dominated. And so this process of elimination can be continued. Second, a pure strategy may be dominated by a mixture of other pure strategies without being dominated by any of them independently. To see this, consider the game in Figure 3.3.

Column R can be eliminated, since it is dominated by, for example, column L. We are left with the reduced game in Figure 3.4.

In this game M is dominated by neither U nor D, but it is dominated by the mixed strategy that selects either U or D with equal probability. (Note, however, that it was not dominated before the elimination of the R column.) And so we are left with the maximally reduced game in Figure 3.5.

This yields us a solution concept: the set of all strategy profiles that assign zero probability to playing any action that would be removed through iterated removal of strictly dominated strategies. Note that this is a much weaker solution concept than Nash equilibrium—the set of strategy profiles will include all the Nash equilibria, but it will include many other

	L	C
U	$3, 1$	$0, 1$
M	$1, 1$	$1, 1$
D	$0, 1$	$4, 1$

FIGURE 3.4: The game from Figure 3.3 after removing the dominated strategy R.

mixed strategies as well. In some games, it will be equal to S, the set of all possible mixed strategies.

Since iterated removal of strictly dominated strategies preserves Nash equilibria, we can use this technique to computational advantage. In the previous example, rather than computing the Nash equilibria in the original 3×3 game, we can now compute them in this 2×2 game, applying the technique described earlier. In some cases, the procedure ends with a single cell; this is the case, for example, with the Prisoner's Dilemma game. In this case we say that the game is *solvable* by iterated elimination.

Clearly, in any finite game, iterated elimination ends after a finite number of iterations. One might worry that, in general, the order of elimination might affect the final outcome. It turns out that this elimination order does not matter when we remove *strictly* dominated strategies. (This is called a *Church–Rosser* property.) However, the elimination order can make a difference to the final reduced game if we remove weakly or very weakly dominated strategies.

Which flavor of domination should we concern ourselves with? In fact, each flavor has advantages and disadvantages, which is why we present all of them here. Strict domination

	L	C
U	$3, 1$	$0, 1$
D	$0, 1$	$4, 1$

FIGURE 3.5: The game from Figure 3.4 after removing the dominated strategy M.

leads to better-behaved iterated elimination: it yields a reduced game which is independent of the elimination order, and iterated elimination is more computationally manageable. There is also a further related advantage that we will defer to Section 3.4. Weak domination can yield smaller reduced games, but under iterated elimination the reduced game can depend on the elimination order. Very weak domination can yield even smaller reduced games, but again these reduced games depend on elimination order. Furthermore, very weak domination does not impose a strict order on strategies: when two strategies are equivalent, each very weakly dominates the other. For this reason, this last form of domination is generally considered the least important.

3.4 RATIONALIZABILITY

A strategy is *rationalizable* if a perfectly rational player could justifiably play it against one or more perfectly rational opponents. Informally, a strategy profile for player i is rationalizable if it is a best response to some beliefs that i could have about the strategies that the other players will take. The wrinkle, however, is that i cannot have arbitrary beliefs about the other players' actions—his beliefs must take into account his knowledge of *their* rationality, which incorporates their knowledge of *his* rationality, their knowledge of his knowledge of their rationality, and so on in an infinite regress. A rationalizable strategy profile is a strategy profile that consists only of rationalizable strategies.

For example, in the Matching Pennies game given in Figure 1.4, the pure strategy *heads* is rationalizable for the row player. First, the strategy *heads* is a best response to the pure strategy *heads* by the column player. Second, believing that the column player would also play *heads* is consistent with the column player's rationality: the column player could believe that the row player would play *tails*, to which the column player's best response is *heads*. It would be rational for the column player to believe that the row player would play *tails* because the column player could believe that the row player believed that the column player would play *tails*, to which *tails* is a best response. Arguing in the same way, we can make our way up the chain of beliefs.

However, not every strategy can be justified in this way. For example, considering the Prisoner's Dilemma game given in Figure 1.1, the strategy C is not rationalizable for the row player, because C is not a best response to any strategy that the column player could play. Similarly, consider the game from Figure 3.3. M is not a rationalizable strategy for the row player: although it *is* a best response to a strategy of the column player's (R), there do not exist any beliefs that the column player could hold about the row player's strategy to which R would be a best response.

Because of the infinite regress, the formal definition of rationalizability is somewhat involved; however, it turns out that there are some intuitive things that we can say about rationalizable strategies. First, Nash equilibrium strategies are always rationalizable: thus, the set of rationalizable strategies (and strategy profiles) is always nonempty. Second, in two-player games rationalizable strategies have a simple characterization: they are those strategies that survive the iterated elimination of strictly dominated strategies. In n-player games there exist strategies which survive iterated removal of dominated strategies but are not rationalizable. In this more general case, rationalizable strategies are those strategies which survive iterative removal of strategies that are never a best response to any strategy profile by the other players.

We now define rationalizability more formally. First we will define an infinite sequence of (possibly mixed) strategies $S_i^0, S_i^1, S_i^2, \ldots$ for each player i. Let $S_i^0 = S_i$; thus, for each agent i, the first element in the sequence is the set of all i's mixed strategies. Let $CH(S)$ denote the convex hull of a set S: the smallest convex set containing all the elements of S. Now we define S_i^k as the set of all strategies $s_i \in S_i^{k-1}$ for which there exists some $s_{-i} \in \prod_{j \neq i} CH(S_j^{k-1})$ such that for all $s_i' \in S_i^{k-1}$, $u_i(s_i, s_{-i}) \geq u_i(s_i', s_{-i})$. That is, a strategy belongs to S_i^k if there is some strategy s_{-i} for the other players in response to which s_i is at least as good as any other strategy from S_i^{k-1}. The convex hull operation allows i to best respond to uncertain beliefs about which strategies from S_j^{k-1} another player j will adopt. $CH(S_j^{k-1})$ is used instead of $\Pi(S_j^{k-1})$, the set of all probability distributions over S_j^{k-1}, because the latter would allow consideration of mixed strategies that are dominated by some pure strategies for j. Player i could not believe that j would play such a strategy because such a belief would be inconsistent with i's knowledge of j's rationality.

Now we define the set of rationalizable strategies for player i as the intersection of the sets $S_i^0, S_i^1, S_i^2, \ldots$.

Definition 3.4.1 (Rationalizable strategies). *The* rationalizable strategies *for player i are* $\bigcap_{k=0}^{\infty} S_i^k$.

3.5 CORRELATED EQUILIBRIUM

The *correlated equilibrium* is a solution concept which generalizes the Nash equilibrium. Some people feel that this is the most fundamental solution concept of all.[3]

In a standard game, each player mixes his pure strategies independently. For example, consider again the Battle of the Sexes game (reproduced here as Figure 3.6) and its mixed-strategy equilibrium.

As we saw in Section 2.3, this game's unique mixed-strategy equilibrium yields each player an expected payoff of 2/3. But now imagine that the two players can observe the result

[3]One Nobel-prize-winning game theorist, R. Myerson, has gone so far as to say that "if there is intelligent life on other planets, in a majority of them, they would have discovered correlated equilibrium before Nash equilibrium."

	LW	WL
LW	2, 1	0, 0
WL	0, 0	1, 2

FIGURE 3.6: Battle of the Sexes game.

of a fair coin flip and can condition their strategies based on that outcome. They can now adopt strategies from a richer set; for example, they could choose "WL if heads, LW if tails." Indeed, this pair forms an equilibrium in this richer strategy space; given that one player plays the strategy, the other player only loses by adopting another. Furthermore, the expected payoff to each player in this so-called correlated equilibrium is $.5 * 2 + .5 * 1 = 1.5$. Thus both agents receive higher utility than they do under the mixed-strategy equilibrium in the uncorrelated case (which had expected payoff of 2/3 for both agents), and the outcome is fairer than either of the pure-strategy equilibria in the sense that the worst-off player achieves higher expected utility. Correlating devices can thus be quite useful.

The aforementioned example had both players observe the exact outcome of the coin flip, but the general setting does not require this. Generally, the setting includes some random variable (the "external event") with a commonly known probability distribution, and a private signal to each player about the instantiation of the random variable. A player's signal can be correlated with the random variable's value and with the signals received by other players, without uniquely identifying any of them. Standard games can be viewed as the degenerate case in which the signals of the different agents are probabilistically independent.

To model this formally, consider n random variables, with a joint distribution over these variables. Imagine that nature chooses according to this distribution, but reveals to each agent only the realized value of his variable, and that the agent can condition his action on this value.[4]

Definition 3.5.1 (Correlated equilibrium). *Given an n-agent game $G = (N, A, u)$, a* correlated equilibrium *is a tuple (v, π, σ), where v is a tuple of random variables $v = (v_1, \ldots, v_n)$ with respective domains $D = (D_1, \ldots, D_n)$, π is a joint distribution over v, $\sigma = (\sigma_1, \ldots, \sigma_n)$ is a vector of mappings $\sigma_i : D_i \mapsto A_i$, and for each agent i and every mapping $\sigma_i' : D_i \mapsto A_i$ it is the case that*

$$\sum_{d \in D} \pi(d) u_i\left(\sigma_1(d_1), \ldots, \sigma_n(d_n)\right) \geq \sum_{d \in D} \pi(d) u_i\left(\sigma_1'(d_1), \ldots, \sigma_n'(d_n)\right).$$

[4]This construction is closely related to one used later in the book in connection with Bayesian Games in Chapter 7.

Note that the mapping is to an action, that is to a pure strategy rather than a mixed one. One could allow a mapping to mixed strategies, but that would add no greater generality. (Do you see why?)

Clearly, for every Nash equilibrium, we can construct an equivalent correlated equilibrium, in the sense that they induce the same distribution on outcomes.

Theorem 3.5.2. *For every Nash equilibrium σ^* there exists a corresponding correlated equilibrium σ.*

The proof is straightforward. Roughly, we can construct a correlated equilibrium from a given Nash equilibrium by letting each $D_i = A_i$ and letting the joint probability distribution be $\pi(d) = \prod_{i \in N} \sigma_i^*(d_i)$. Then we choose σ_i as the mapping from each d_i to the corresponding a_i. When the agents play the strategy profile σ, the distribution over outcomes is identical to that under σ^*. Because the v_i's are uncorrelated and no agent can benefit by deviating from σ^*, σ is a correlated equilibrium.

On the other hand, not every correlated equilibrium is equivalent to a Nash equilibrium; the Battle-of-the-Sexes example given earlier provides a counter-example. Thus, correlated equilibrium is a strictly weaker notion than Nash equilibrium.

Finally, we note that correlated equilibria can be combined together to form new correlated equilibria. Thus, if the set of correlated equilibria of a game G does not contain a single element, it is infinite. Indeed, any convex combination of correlated equilibrium payoffs can itself be realized as the payoff profile of some correlated equilibrium. The easiest way to understand this claim is to imagine a public random device that selects which of the correlated equilibria will be played; next, another random number is chosen in order to allow the chosen equilibrium to be played. Overall, each agent's expected payoff is the weighted sum of the payoffs from the correlated equilibria that were combined. Since no agent has an incentive to deviate regardless of the probabilities governing the first random device, we can achieve any convex combination of correlated equilibrium payoffs. Finally, observe that having two stages of random number generation is not necessary: we can simply derive new domains D and a new joint probability distribution π from the D's and π's of the original correlated equilibria, and so perform the random number generation in one step.

3.6 TREMBLING-HAND PERFECT EQUILIBRIUM

Another important solution concept is the *trembling-hand perfect equilibrium*, or simply *perfect equilibrium*. While rationalizability is a weaker notion than that of a Nash equilibrium, perfection is a stronger one. Several equivalent definitions of the concept exist. In the following definition, recall that a fully mixed strategy is one that assigns every action a strictly positive probability.

Definition 3.6.1 (Trembling-hand perfect equilibrium). *A mixed strategy S is a* (trembling-hand) perfect equilibrium *of a normal-form game G if there exists a sequence $S^0, S^1, \ldots$ of fully mixed-strategy profiles such that* $\lim_{n\to\infty} S^n = S$, *and such that for each S^k in the sequence and each player i, the strategy s_i is a best response to the strategies s^k_{-i}.*

Perfect equilibria are an involved topic, and relate to other subtle refinements of the Nash equilibrium such as the *proper equilibrium*. The notes at the end of the booklet point the reader to further readings on this topic. We should, however, at least explain the term "trembling hand." One way to think about the concept is as requiring that the equilibrium be robust against slight errors—"trembles"—on the part of players. In other words, one's action ought to be the best response not only against the opponents' equilibrium strategies, but also against small perturbation of those. However, since the mathematical definition speaks about arbitrarily small perturbations, whether these trembles in fact model player fallibility or are merely a mathematical device is open to debate.

3.7 ϵ-NASH EQUILIBRIUM

Another solution concept reflects the idea that players might not care about changing their strategies to a best response when the amount of utility that they could gain by doing so is very small. This leads us to the idea of an ϵ-Nash equilibrium.

Definition 3.7.1 (ϵ-Nash). *Fix $\epsilon > 0$. A strategy profile $s = (s_1, \ldots, s_n)$ is an ϵ-Nash equilibrium if, for all agents i and for all strategies $s'_i \neq s_i$, $u_i(s_i, s_{-i}) \geq u_i(s'_i, s_{-i}) - \epsilon$.*

This concept has various attractive properties. ϵ-Nash equilibria always exist; indeed, every Nash equilibrium is surrounded by a region of ϵ-Nash equilibria for any $\epsilon > 0$. The argument that agents are indifferent to sufficiently small gains is convincing to many. Further, the concept can be computationally useful: algorithms that aim to identify ϵ-Nash equilibria need to consider only a finite set of mixed-strategy profiles rather than the whole continuous space. (Of course, the size of this finite set depends on both ϵ and on the game's payoffs.) Since computers generally represent real numbers using a floating-point approximation, it is usually the case that even methods for the "exact" computation of Nash equilibria actually find only ϵ-equilibria where ϵ is roughly the "machine precision" (on the order of 10^{-16} or less for most modern computers). ϵ-Nash equilibria are also important to multiagent learning algorithms, not discussed in this booklet.

However, ϵ-Nash equilibria also have several drawbacks. First, although Nash equilibria are always surrounded by ϵ-Nash equilibria, the reverse is not true. Thus, a given ϵ-Nash equilibrium is not necessarily close to any Nash equilibrium. This undermines the sense in

	L	R
U	$1,1$	$0,0$
D	$1+\frac{\epsilon}{2},1$	$500,500$

FIGURE 3.7: A game with an interesting ϵ-Nash equilibrium.

which ϵ-Nash equilibria can be understood as approximations of Nash equilibria. Consider the game in Figure 3.7.

This game has a unique Nash equilibrium of (D, R), which can be identified through the iterated removal of dominated strategies. (D dominates U for player 1; on the removal of U, R dominates L for player 2.) (D, R) is also an ϵ-Nash equilibrium, of course. However, there is also another ϵ-Nash equilibrium: (U, L). This game illustrates two things.

First, neither player's payoff under the ϵ-Nash equilibrium is within ϵ of his payoff in a Nash equilibrium; indeed, in general both players' payoffs under an ϵ-Nash equilibrium can be arbitrarily less than in any Nash equilibrium. The problem is that the requirement that player 1 cannot gain more than ϵ by deviating from the ϵ-Nash equilibrium strategy profile of (U, L) does not imply that *player 2* would not be able to gain more than ϵ by best responding to player 1's deviation.

Second, some ϵ-Nash equilibria might be very unlikely to arise in play. Although player 1 might not care about a gain of $\frac{\epsilon}{2}$, he might reason that the fact that D dominates U would lead player 2 to expect him to play D, and that player 2 would thus play R in response. Player 1 might thus play D because it is his best response to R. Overall, the idea of ϵ-approximation is much messier when applied to the identification of a fixed point than when it is applied to a (single-objective) optimization problem.

3.8 EVOLUTIONARILY STABLE STRATEGIES

Roughly speaking, an evolutionarily stable strategy is a mixed strategy that is "resistant to invasion" by new strategies. As can be gleaned from the name, inspiration for to concept of evolutionarily stable strategies comes from evolutionary biology. There one speaks about different species within a population, and how each species' relative "fitness" causes its proportion within the population to grow or shrink. In our setting the species are those agents playing a particular strategy. Then suppose that a small population of "invaders" playing a different strategy is added to the population. The original strategy is considered to be an ESS if it gets a higher payoff against the resulting mixture of the new and old strategies than the invaders do, thereby "chasing out" the invaders.

More formally, we have the following.

Definition 3.8.1 (Evolutionarily stable strategy (ESS)). *Given a symmetric two-player normal-form game $G = (\{1, 2\}, A, u)$ and a mixed strategy S, we say that S is an evolutionarily stable strategy if and only if for some $\epsilon > 0$ and for all other strategies S' it is the case that*

$$u(S, (1-\epsilon)S + \epsilon S') > u(S', (1-\epsilon)S + \epsilon S').$$

We can use properties of expectation to state this condition equivalently as

$$(1-\epsilon)u(S, S) + \epsilon u(S, S') > (1-\epsilon)u(S', S) + \epsilon u(S', S').$$

Note that, since this only need hold for small ϵ, this is equivalent to requiring that either $u(S, S) > u(S', S)$ holds, or else both $u(S, S) = u(S', S)$ and $u(S, S') > u(S', S')$ hold. Note that this is a strict definition. We can also state a weaker definition of ESS.

Definition 3.8.2 (Weak ESS). *S is a* weak evolutionarily stable strategy *if and only if for some $\epsilon > 0$ and for all S' it is the case that either $u(S, S) > u(S', S)$ holds, or else both $u(S, S) = u(S', S)$ and $u(S, S') \geq u(S', S')$ hold.*

This weaker definition includes strategies in which the invader does just as well against the original population as it does against itself. In these cases the population using the invading strategy will not grow, but it will also not shrink.

We illustrate the concept of ESS with the instance of the *Hawk–Dove* game shown in Figure 3.8. The story behind this game might be as follows. Two animals are fighting over a prize such as a piece of food. Each animal can choose between two behaviors: an aggressive hawkish behavior H, or an accommodating dovish behavior D. The prize is worth 6 to each of them. Fighting costs each player 5. When a hawk meets a dove he gets the prize without a fight, and hence the payoffs are 6 and 0, respectively. When two doves meet they split the prize

	H	D
H	$-2, -2$	$6, 0$
D	$0, 6$	$3, 3$

FIGURE 3.8: Hawk–Dove game.

without a fight, hence a payoff of 3 to each one. When two hawks meet a fight breaks out, costing each player 5 (or, equivalently, yielding −5). In addition, each player has a 50% chance of ending up with the prize, adding an expected benefit of 3, for an overall payoff of −2.

It is not hard to verify that the game has a unique symmetric Nash equilibrium (S, S), where $S = (\frac{3}{5}, \frac{2}{5})$, and that S is also the unique ESS of the game. To confirm that S is an ESS, we need that for all $S' \neq S$, $u(S, S) = u(S', S)$ and $u(S, S') > u(S', S')$. The equality condition is true of any mixed strategy equilibrium with full support, so follows directly. To demonstrate that the inequality holds, it is sufficient to find the S'—or equivalently, the probability of playing H—that minimizes $f(S') = u(S, S') - u(S', S')$. Expanding $f(S')$ we see that it is a quadratic equation with the (unique) maximum $S' = S$, proving our result.

The connection between an ESS and a Nash equilibrium is not accidental. The following two theorems capture this connection.

Theorem 3.8.3. *Given a symmetric two-player normal-form game $G = (\{1, 2\}, A, u)$ and a mixed strategy S, if S is an evolutionarily stable strategy then (S, S) is a Nash equilibrium of G.*

This is easy to show. Note that by definition an ESS S must satisfy

$$u(S, S) \geq u(S', S).$$

In other words, it is a best response to itself and thus must be a Nash equilibrium. However, not every Nash equilibrium is an ESS; this property is guaranteed only for strict equilibria.

Theorem 3.8.4. *Given a symmetric two-player normal-form game $G = (\{1, 2\}, A, u)$ and a mixed strategy S, if (S, S) is a strict (symmetric) Nash equilibrium of G, then S is an evolutionarily stable strategy.*

This is also easy to show. Note that for any strict Nash equilibrium S it must be the case that

$$u(S, S) > u(S', S).$$

But this satisfies the first criterion of an ESS.

CHAPTER 4

Games With Sequential Actions: The Perfect-Information Extensive Form

In Chapter 1 we assumed that a game is represented in normal form: effectively, as a big table. In some sense, this is reasonable. The normal form is conceptually straightforward, and most game theorists see it as fundamental. While many other representations exist to describe finite games, we will see in this chapter and in those that follow that each of them has an "induced normal form": a corresponding normal-form representation that preserves game-theoretic properties such as Nash equilibria. Thus the results given in previous chapters hold for all finite games, no matter how they are represented; in that sense the normal-form representation is universal.

In this chapter we will look at perfect-information extensive-form games, a finite representation that does not always assume that players act simultaneously. This representation is in general exponentially smaller than its induced normal form, and furthermore can be much more natural to reason about. While the Nash equilibria of an extensive-form game can be found through its induced normal form, computational benefit can be had by working with the extensive form directly. Furthermore, there are other solution concepts, such as subgame-perfect equilibrium (see Section 4.3), which explicitly refer to the sequence in which players act and which are therefore not meaningful when applied to normal-form games.

The normal-form game representation does not incorporate any notion of sequence, or time, of the actions of the players. The *extensive (*or *tree) form* is an alternative representation that makes the temporal structure explicit. In this chapter we discuss the special case of *perfect information* extensive-form games.We will restrict the discussion to finite games, that is, to games represented as finite trees.

4.1 DEFINITION

Informally speaking, a perfect-information game in extensive form (or, more simply, a perfect-information game) is a tree in the sense of graph theory, in which each node represents the choice of one of the players, each edge represents a possible action, and the leaves represent final outcomes over which each player has a utility function. Indeed, in certain circles (in particular, in artificial intelligence), these are known simply as game trees. Formally, we define them as follows.

Definition 4.1.1 (Perfect-information game). *A (finite) perfect-information game (in extensive form) is a tuple* $G = (N, A, H, Z, \chi, \rho, \sigma, u)$*, where:*

- N *is a set of* n *players;*
- A *is a (single) set of actions;*
- H *is a set of nonterminal choice nodes;*
- Z *is a set of terminal nodes, disjoint from* H*;*
- $\chi : H \mapsto 2^A$ *is the action function, which assigns to each choice node a set of possible actions;*
- $\rho : H \mapsto N$ *is the player function, which assigns to each nonterminal node a player* $i \in N$ *who chooses an action at that node;*
- $\sigma : H \times A \mapsto H \cup Z$ *is the successor function, which maps a choice node and an action to a new choice node or terminal node such that for all* $h_1, h_2 \in H$ *and* $a_1, a_2 \in A$*, if* $\sigma(h_1, a_1) = \sigma(h_2, a_2)$ *then* $h_1 = h_2$ *and* $a_1 = a_2$*; and*
- $u = (u_1, \ldots, u_n)$*, where* $u_i : Z \mapsto \mathbb{R}$ *is a real-valued utility function for player* i *on the terminal nodes* Z*.*

Since the choice nodes form a tree, we can unambiguously identify a node with its *history*, that is, the sequence of choices leading from the root node to it. We can also define the *descendants* of a node h, namely all the choice and terminal nodes in the subtree rooted in h.

An example of such a game is the *Sharing game*. Imagine a brother and sister following the following protocol for sharing two indivisible and identical presents from their parents. First the brother suggests a split, which can be one of three—he keeps both, she keeps both, or they each keep one. Then the sister chooses whether to accept or reject the split. If she accepts they each get their allocated present(s), and otherwise neither gets any gift. Assuming both siblings value the two presents equally and additively, the tree representation of this game is shown in Figure 4.1.

4.2 STRATEGIES AND EQUILIBRIA

A pure strategy for a player in a perfect-information game is a complete specification of which deterministic action to take at every node belonging to that player. A more formal definition follows.

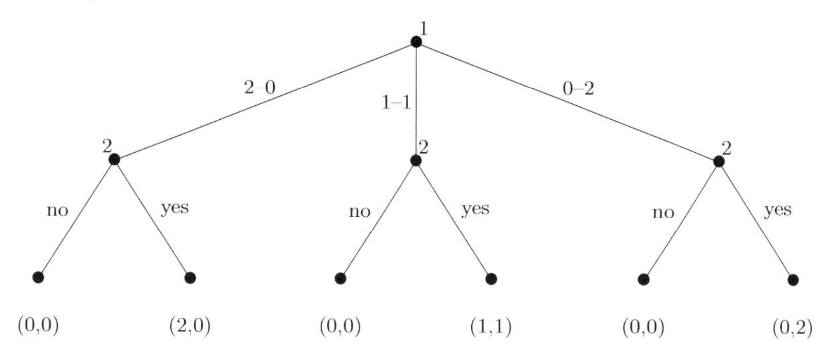

FIGURE 4.1: The Sharing game.

Definition 4.2.1 (Pure strategies). *Let $G = (N, A, H, Z, \chi, \rho, \sigma, u)$ be a perfect-information extensive-form game. Then the pure strategies of player i consist of the Cartesian product $\prod_{h \in H, \rho(h)=i} \chi(h)$.*

Notice that the definition contains a subtlety. An agent's strategy requires a decision at each choice node, regardless of whether or not it is possible to reach that node given the other choice nodes. In the Sharing game above the situation is straightforward—player 1 has three pure strategies, and player 2 has eight, as follows.

$S_1 = \{2\text{–}0, 1\text{–}1, 0\text{–}2\}$
$S_2 = \{(yes, yes, yes), \ (yes, yes, no), \ (yes, no, yes), \ (yes, no, no), \ (no, yes, yes), (no, yes, no), (no, no, yes), (no, no, no)\}$

But now consider the game shown in Figure 4.2.

In order to define a complete strategy for this game, each of the players must choose an action at each of his two choice nodes. Thus we can enumerate the pure strategies of the players

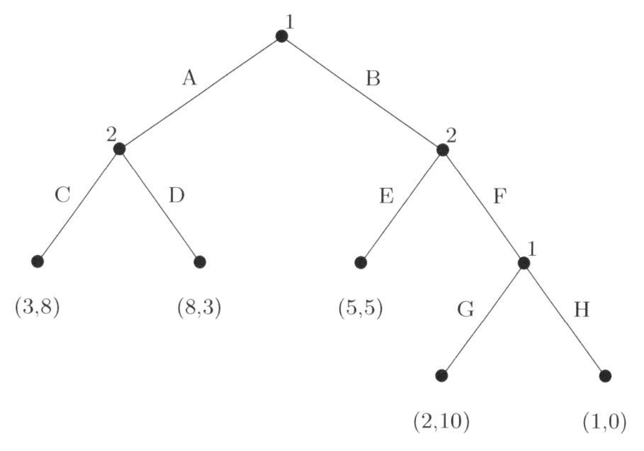

FIGURE 4.2: A perfect-information game in extensive form.

	(C,E)	(C,F)	(D,E)	(D,F)
(A,G)	3,8	3,8	8,3	8,3
(A,H)	3,8	3,8	8,3	8,3
(B,G)	5,5	2,10	5,5	2,10
(B,H)	5,5	1,0	5,5	1,0

FIGURE 4.3: The game from Figure 4.2 in normal form.

as follows.

$$S_1 = \{(A, G), (A, H), (B, G), (B, H)\}$$
$$S_2 = \{(C, E), (C, F), (D, E), (D, F)\}$$

It is important to note that we have to include the strategies (A, G) and (A, H), even though once player 1 has chosen A then his own G-versus-H choice is moot.

The definition of best response and Nash equilibria in this game are exactly as they are for normal-form games. Indeed, this example illustrates how every perfect-information game can be converted to an equivalent normal-form game. For example, the perfect-information game of Figure 4.2 can be converted into the normal form image of the game, shown in Figure 4.3. Clearly, the strategy spaces of the two games are the same, as are the pure-strategy Nash equilibria. (Indeed, both the mixed strategies and the mixed-strategy Nash equilibria of the two games are also the same; however, we defer further discussion of mixed strategies until we consider imperfect-information games in Chapter 5.)

In this way, for every perfect-information game there exists a corresponding normal-form game. Note, however, that the temporal structure of the extensive-form representation can result in a certain redundancy within the normal form. For example, in Figure 4.3 there are 16 different outcomes, while in Figure 4.2 there are only 5; the payoff $(3, 8)$ occurs only once in Figure 4.2 but four times in Figure 4.3. One general lesson is that while this transformation can always be performed, it can result in an exponential blowup of the game representation.

This is an important lesson, since the didactic examples of normal-form games are very small, wrongly suggesting that this form is more compact.

The normal form gets its revenge, however, since the reverse transformation—from the normal form to the perfect-information extensive form—does not always exist. Consider, for example, the Prisoner's Dilemma game from Figure 1.1. A little experimentation will convince the reader that there does not exist a perfect-information game that is equivalent in the sense of having the same strategy profiles and the same payoffs. Intuitively, the problem is that perfect-information extensive-form games cannot model simultaneity. The general characterization of the class of normal-form games for which there exist corresponding perfect-information games in extensive form is somewhat complex.

The reader will have noticed that we have so far concentrated on pure strategies and pure Nash equilibria in extensive-form games. There are two reasons for this, or perhaps one reason and one excuse. The reason is that mixed strategies introduce a new subtlety, and it is convenient to postpone discussion of it. The excuse (which also allows the postponement, though not for long) is the following theorem.

Theorem 4.2.2. *Every (finite) perfect-information game in extensive form has a pure-strategy Nash equilibrium.*

This is perhaps the earliest result in game theory, due to Zermelo in 1913 (see the historical notes at the end of the book). The intuition here should be clear; since players take turns, and everyone gets to see everything that happened thus far before making a move, it is never necessary to introduce randomness into action selection in order to find an equilibrium. We will see this plainly when we discuss *backward induction* below. Both this intuition and the theorem will cease to hold when we discuss more general classes of games such as imperfect-information games in extensive form. First, however, we discuss an important refinement of the concept of Nash equilibrium.

4.3 SUBGAME-PERFECT EQUILIBRIUM

As we have discussed, the notion of Nash equilibrium is as well defined in perfect-information games in extensive form as it is in the normal form. However, as the following example shows, the Nash equilibrium can be too weak a notion for the extensive form. Consider again the perfect-information extensive-form game shown in Figure 4.2. There are three pure-strategy Nash equilibria in this game: $\{(A, G), (C, F)\}$, $\{(A, H), (C, F)\}$, and $\{(B, H), (C, E)\}$. This can be determined by examining the normal form image of the game, as indicated in Figure 4.4.

However, examining the normal form image of an extensive-form game obscures the game's temporal nature. To illustrate a problem that can arise in certain equilibria of

	(C,E)	(C,F)	(D,E)	(D,F)
(A,G)	3,8	3,8	8,3	8,3
(A,H)	3,8	3,8	8,3	8,3
(B,G)	5,5	2,10	5,5	2,10
(B,H)	5,5	1,0	5,5	1,0

FIGURE 4.4: Equilibria of the game from Figure 4.2.

extensive-form games, in Figure 4.5 we contrast the equilibria $\{(A, G), (C, F)\}$ and $\{(B, H), (C, E)\}$ by drawing them on the extensive-form game tree.

First consider the equilibrium $\{(A, G), (C, F)\}$. If player 1 chooses A then player 2 receives a higher payoff by choosing C than by choosing D. If player 2 played the strategy (C, E) rather than (C, F) then player 1 would prefer to play B at the first node in the tree; as it is, player 1 gets a payoff of 3 by playing A rather than a payoff of 2 by playing B. Hence we have an equilibrium.

The second equilibrium $\{(B, H), (C, E)\}$ is less intuitive. First, note that $\{(B, G), (C, E)\}$ is *not* an equilibrium: player 2's best response to (B, G) is (C, F). Thus, the only reason that player 2 chooses to play the action E is that he knows that player 1 would play H at his second decision node. This behavior by player 1 is called a *threat*: by committing to choose an action that is harmful to player 2 in his second decision node, player 1 can cause player 2 to avoid that part of the tree. (Note that player 1 benefits from making this threat: he gets a payoff of 5 instead of 2 by playing (B, H) instead of (B, G).) So far so good. The problem, however, is that player 2 may not consider player 1's threat to be credible: if player 1 did reach his final decision node, actually choosing H over G would also reduce player 1's own utility. If player 2 played F, would player 1 really follow through on his threat and play H, or would he relent and pick G instead?

To formally capture the reason why the $\{(B, H), (C, E)\}$ equilibrium is unsatisfying, and to define an equilibrium refinement concept that does not suffer from this problem, we first define the notion of a subgame.

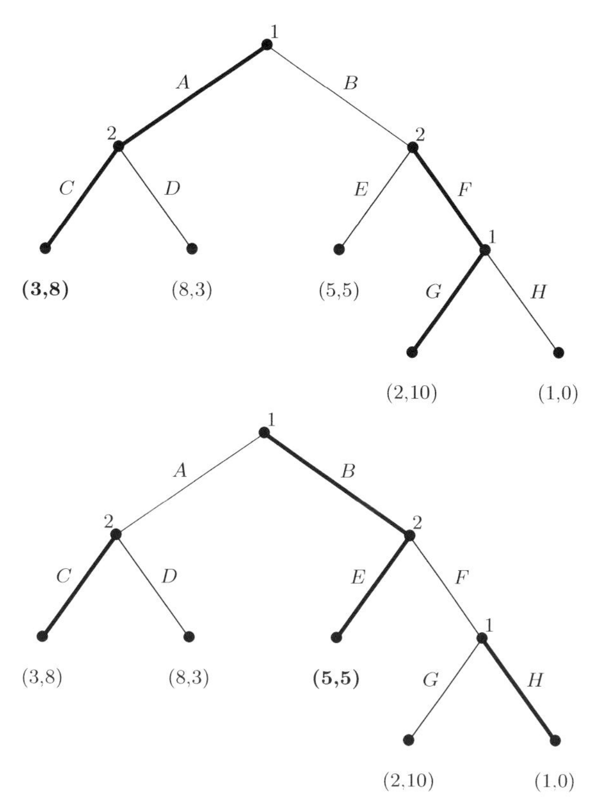

FIGURE 4.5: Two out of the three equilibria of the game from Figure 4.2: $\{(A, G), (C, F)\}$ and $\{(B, H), (C, E)\}$. Bold edges indicate players' choices at each node.

Definition 4.3.1 (Subgame). *Given a perfect-information extensive-form game G, the* subgame *of G rooted at node h is the restriction of G to the descendants of h. The set of subgames of G consists of all of subgames of G rooted at some node in G.*

Now we can define the notion of a *subgame-perfect equilibrium*, a refinement of the Nash equilibrium in perfect-information games in extensive form, which eliminates those unwanted Nash equilibria.[1]

Definition 4.3.2 (Subgame-perfect equilibrium). *The* subgame-perfect equilibria *(SPE) of a game G are all strategy profiles s such that for any subgame G' of G, the restriction of s to G' is a Nash equilibrium of G'.*

Since G is in particular its own subgame, every SPE is also a Nash equilibrium. Furthermore, although SPE is a stronger concept than Nash equilibrium (i.e., every SPE is a NE, but

[1]Note that the word "perfect" is used in two different senses here.

not every NE is a SPE) it is still the case that every perfect-information extensive-form game has at least one subgame-perfect equilibrium.

This definition rules out "noncredible threats," of the sort illustrated in the above example. In particular, note that the extensive-form game in Figure 4.2 has only one subgame-perfect equilibrium, $\{(A, G), (C, F)\}$. Neither of the other Nash equilibria is subgame perfect. Consider the subgame rooted in player 1's second choice node. The unique Nash equilibrium of this (trivial) game is for player 1 to play G. Thus the action H, the restriction of the strategies (A, H) and (B, H) to this subgame, is not optimal in this subgame, and cannot be part of a subgame-perfect equilibrium of the larger game.

4.4 BACKWARD INDUCTION

Inherent in the concept of subgame-perfect equilibrium is the principle of *backward induction*. One identifies the equilibria in the "bottom-most" subgame trees, and assumes that those equilibria will be played as one backs up and considers increasingly larger trees. We can use this procedure to compute a sample Nash equilibrium. This is good news: not only are we guaranteed to find a subgame-perfect equilibrium (rather than possibly finding a Nash equilibrium that involves noncredible threats), but also this procedure is computationally simple.

The algorithm BACKWARDINDUCTION is described in Figure 4.6. The variable $util\_at\_child$ is a vector denoting the utility for each player at the child node; $util\_at\_child_{\rho(h)}$ denotes the element of this vector corresponding to the utility for player $\rho(h)$ (the player who gets to move at node h). Similarly, $best\_util$ is a vector giving utilities for each player.

Observe that this procedure does not return an equilibrium strategy for each of the n players, but rather describes how to label each node with a vector of n real numbers. This

function BACKWARDINDUCTION (node h) **returns** $u(h)$
if $h \in Z$ **then**
  return $u(h)$ // h is a terminal node
$best\_util \leftarrow -\infty$
forall $a \in \chi(h)$ **do**
  $util\_at\_child \leftarrow$ BACKWARDINDUCTION($\sigma(h, a)$)
  if $util\_at\_child_{\rho(h)} > best\_util_{\rho(h)}$ **then**
    $best\_util \leftarrow util\_at\_child$
return $best\_util$

FIGURE 4.6: Procedure for finding the value of a sample (subgame-perfect) Nash equilibrium of a perfect-information extensive-form game.

labeling can be seen as an extension of the game's utility function to the nonterminal nodes H. The players' equilibrium strategies follow straightforwardly from this extended utility function: every time a given player i has the opportunity to act at a given node $h \in H$ (i.e., $\rho(h) = i$), that player will choose an action $a_i \in \chi(h)$ that solves $\arg\max_{a_i \in \chi(h)} u_i(\sigma(a_i, h))$. These strategies can also be returned by BackwardInduction given some extra bookkeeping.

In general in this booklet we do not address computational issues, so this example could be misleading without additional explanation. While the procedure demonstrates that in principle a sample SPE is effectively computable, in practice many game trees are not enumerated in advance and are hence unavailable for backward induction. For example, the extensive-form representation of chess has around 10^{150} nodes, which is vastly too large to represent explicitly.

We note that BackwardInduction has another name in the two-player, zero-sum context: the *minimax algorithm*. Recall that in such games, only a single payoff number is required to characterize any outcome. Player 1 wants to maximize this number, while player 2 wants to minimize it. In this context BackwardInduction can be understood as propagating these single payoff numbers from the root of the tree up to the root. Each decision node for player 1 is labeled with the maximum of the labels of its child nodes (representing the fact that player 1 would choose the corresponding action), and each decision node for player 2 is labeled with the minimum of that node's children's labels. The label on the root node is the value of the game: player 1's payoff in equilibrium.

As we said, real-world games—even zero-sum ones, such as chess—cannot be represented explicitly. Such games require the gradual development of the tree, and its heuristic search. At least in the context of zero-sum games, considerable effort has gone into such search algorithms. The best-known one, AlphaBetaPruning, is a heuristic version of the minimax algorithm.

Despite the fact that strong arguments can be made in its favor, the concept of backward induction is not without controversy. To see why this is, consider the well-known *Centipede game*, depicted in Figure 4.7. (The game starts at the node at the upper left.) In this game two players alternate in making decisions, at each turn choosing between going "down" and ending the game or going "across" and continuing it (except at the last node where going "across" also ends the game). The payoffs are constructed in such a way that the only SPE is for each player

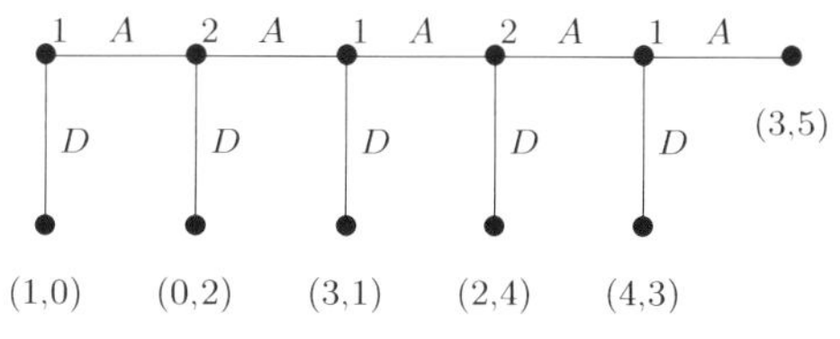

FIGURE 4.7: The Centipede game.

to always choose to go down. So see why, consider the last choice. Clearly at that point the best choice for the player is to go down. Since this is the case, going down is also the best choice for the other player in the previous choice point. By induction the same argument holds for all choice points.

This would seem to be the end of this story, except for two pesky factors. The first problem is that thc SPE prediction in this case flies in the face of intuition. Indeed, in laboratory experiments subjects in fact continue to stay play "across" until close to the end of the game. The second problem is theoretical. Imagine that you are the second player in the game, and in the first step of the game the first player actually goes across. What should you do? The SPE suggests you should go down, but the same analysis suggests that you would not have gotten to this choice point in the first place. In other words, you have reached a state to which your analysis has given a probability of zero. How should you amend your beliefs and course of action based on this measure-zero event? It turns out this seemingly small inconvenience actually raises a fundamental problem in game theory. We will not develop the subject further here, but let us only mention that there exist different accounts of this situation, and they depend on the probabilistic assumptions made, on what is common knowledge (in particular, whether there is common knowledge of rationality), and on exactly how one revises one's beliefs in the face of measure-zero events.

CHAPTER 5

Generalizing the Extensive Form: Imperfect-Information Games

In Chapter 4, in our discussion of extensive-form games we allowed players to specify the action that they would take at every choice node of the game. This implies that players know the node they are in, and—recalling that in such games we equate nodes with the histories that led to them—all the prior choices, including those of other agents. For this reason we have called these *perfect-information games*.

We might not always want to make such a strong assumption about our players and our environment. In many situations we may want to model agents needing to act with partial or no knowledge of the actions taken by others, or even agents with limited memory of their own past actions. The sequencing of choices allows us to represent such ignorance to a limited degree; an "earlier" choice might be interpreted as a choice made without knowing the "later" choices. However, so far we could not represent two choices made in the same play of the game in mutual ignorance of each other.

5.1 DEFINITION

Imperfect-information games in extensive form address this limitation. An imperfect-information game is an extensive-form game in which each player's choice nodes are partitioned into information sets; intuitively, if two choice nodes are in the same information set then the agent cannot distinguish between them.

Definition 5.1.1 (Imperfect-information game). *An imperfect-information game (in extensive form) is a tuple* $(N, A, H, Z, \chi, \rho, \sigma, u, I)$*, where:*

- $(N, A, H, Z, \chi, \rho, \sigma, u)$ *is a perfect-information extensive-form game; and*
- $I = (I_1, \ldots, I_n)$*, where* $I_i = (I_{i,1}, \ldots, I_{i,k_i})$ *is an equivalence relation on (i.e., a partition of)* $\{h \in H : \rho(h) = i\}$ *with the property that* $\chi(h) = \chi(h')$ *and* $\rho(h) = \rho(h')$ *whenever there exists a* j *for which* $h \in I_{i,j}$ *and* $h' \in I_{i,j}$.

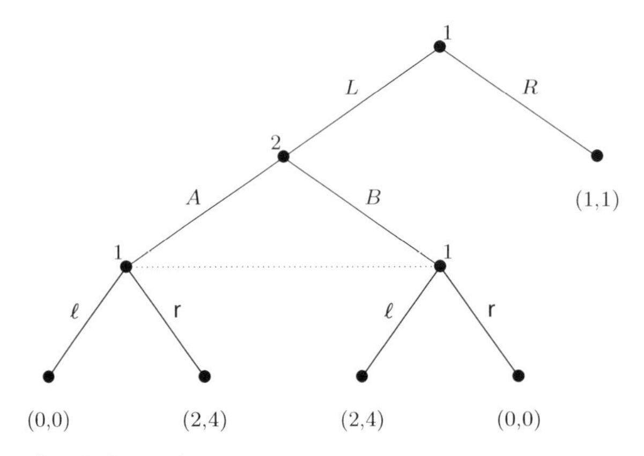

FIGURE 5.1: An imperfect-information game.

Note that in order for the choice nodes to be truly indistinguishable, we require that the set of actions at each choice node in an information set be the same (otherwise, the player would be able to distinguish the nodes). Thus, if $I_{i,j} \in I_i$ is an equivalence class, we can unambiguously use the notation $\chi(I_{i,j})$ to denote the set of actions available to player i at any node in information set $I_{i,j}$.

Consider the imperfect-information extensive-form game shown in Figure 5.1. In this game, player 1 has two information sets: the set including the top choice node, and the set including the bottom choice nodes. Note that the two bottom choice nodes in the second information set have the same set of possible actions. We can regard player 1 as not knowing whether player 2 chose A or B when he makes her choice between ℓ and r.

5.2 STRATEGIES AND EQUILIBRIA

A pure strategy for an agent in an imperfect-information game selects one of the available actions in each information set of that agent.

Definition 5.2.1 (Pure strategies). *Let $G = (N, A, H, Z, \chi, \rho, \sigma, u, I)$ be an imperfect-information extensive-form game. Then the pure strategies of player i consist of the Cartesian product $\prod_{I_{i,j} \in I_i} \chi(I_{i,j})$.*

Thus perfect-information games can be thought of as a special case of imperfect-information games, in which every equivalence class of each partition is a singleton.

Consider again the Prisoner's Dilemma game, shown as a normal-form game in Figure 1.1. An equivalent imperfect-information game in extensive form is given in Figure 5.2.

Note that we could have chosen to make player 2 choose first and player 1 choose second.

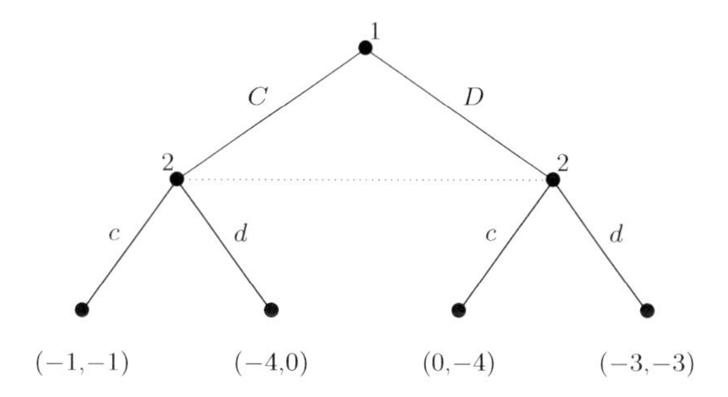

FIGURE 5.2: The Prisoner's Dilemma game in extensive form.

Recall that perfect-information games were not expressive enough to capture the Prisoner's Dilemma game and many other ones. In contrast, as is obvious from this example, any normal-form game can be trivially transformed into an equivalent imperfect-information game. However, this example is also special in that the Prisoner's Dilemma is a game with a dominant strategy solution, and thus in particular a pure-strategy Nash equilibrium. This is not true in general for imperfect-information games. To be precise about the equivalence between a normal-form game and its extensive-form image we must consider mixed strategies, and this is where we encounter a new subtlety.

As we did for perfect-information games, we can define the normal-form game corresponding to any given imperfect-information game; this normal game is again defined by enumerating the pure strategies of each agent. Now, we define the set of mixed strategies of an imperfect-information game as simply the set of mixed strategies in its image normal-form game; in the same way, we can also define the set of Nash equilibria.[1] However, we can also define the set of *behavioral strategies* in the extensive-form game. These are the strategies in which each agent's (potentially probabilistic) choice at each node is made independently of his choices at other nodes. The difference is substantive, and we illustrate it in the special case of perfect-information games. For example, consider the game of Figure 4.2. A strategy for player 1 that selects A with probability .5 and G with probability .3 is a behavioral strategy. In contrast, the mixed strategy $(.6(A, G), .4(B, H))$ is not a behavioral strategy for that player, since the choices made by him at the two nodes are not independent (in fact, they are perfectly correlated).

In general, the expressive power of behavioral strategies and the expressive power of mixed strategies are noncomparable; in some games there are outcomes that are achieved via

[1]Note that we have defined two transformations—one from any normal-form game to an imperfect-information game, and one in the other direction. However the first transformation is not one to one, and so if we transform a normal-form game to an extensive-form one and then back to normal form, we will not in general get back the same game we started out with. However, we will get a game with identical strategy spaces and equilibria.

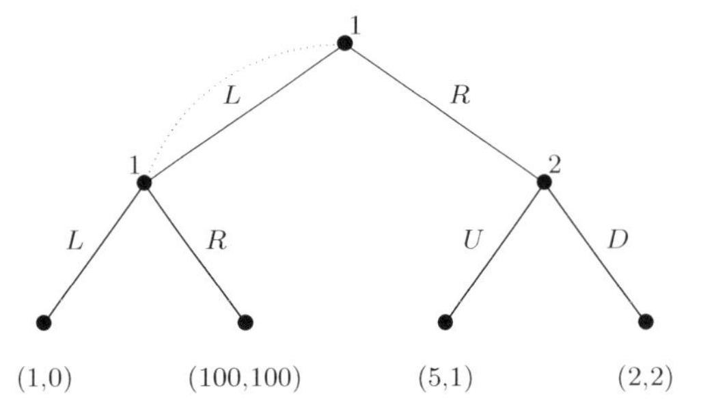

FIGURE 5.3: A game with imperfect recall.

mixed strategies but not any behavioral strategies, and in some games it is the other way around.

Consider for example the game in Figure 5.3. In this game, when considering mixed strategies (but not behavioral strategies), R is a strictly dominant strategy for agent 1, D is agent 2's strict best response, and thus (R, D) is the unique Nash equilibrium. Note in particular that in a mixed strategy, agent 1 decides probabilistically whether to play L or R in his information set, but once he decides he plays that pure strategy consistently. Thus the payoff of 100 is irrelevant in the context of mixed strategies. On the other hand, with behavioral strategies agent 1 gets to randomize afresh each time he finds himself in the information set. Noting that the pure strategy D is weakly dominant for agent 2 (and in fact is the unique best response to all strategies of agent 1 other than the pure strategy L), agent 1 computes the best response to D as follows. If he uses the behavioral strategy $(p, 1-p)$ (i.e., choosing L with probability p each time he finds himself in the information set), his expected payoff is

$$1 * p^2 + 100 * p(1-p) + 2 * (1-p).$$

The expression simplifies to $-99p^2 + 98p + 2$, whose maximum is obtained at $p = 98/198$. Thus $(R, D) = ((0, 1), (0, 1))$ is no longer an equilibrium in behavioral strategies, and instead we get the equilibrium $((98/198, 100/198), (0, 1))$.

There is, however, a broad class of imperfect-information games in which the expressive power of mixed and behavioral strategies coincides. This is the class of games of *perfect recall*. Intuitively speaking, in these games no player forgets any information he knew about moves made so far; in particular, he remembers precisely all his own moves. A formal definition follows.

Definition 5.2.2 (Perfect recall). *Player i has* perfect recall *in an imperfect-information game G if for any two nodes h, h' that are in the same information set for player i, for any path*

$h_0, a_0, h_1, a_1, h_2, \ldots, h_n, a_n, h$ from the root of the game to h (where the h_j are decision nodes and the a_j are actions) and for any path $h_0, a'_0, h'_1, a'_1, h'_2, \ldots, h'_m, a'_m, h'$ from the root to h' it must be the case that:

1. *$n = m$;*
2. *for all $0 \leq j \leq n$, h_j and h'_j are in the same equivalence class for player i and;*
3. *for all $0 \leq j \leq n$, if $\rho(h_j) = i$ (i.e., h_j is a decision node of player i), then $a_j = a'_j$.*

G is a game of perfect recall if every player has perfect recall in it.

Clearly, every perfect-information game is a game of perfect recall.

Theorem 5.2.3 (Kuhn, 1953). *In a game of perfect recall, any mixed strategy of a given agent can be replaced by an equivalent behavioral strategy, and any behavioral strategy can be replaced by an equivalent mixed strategy. Here two strategies are equivalent in the sense that they induce the same probabilities on outcomes, for any fixed strategy profile (mixed or behavioral) of the remaining agents.*

As a corollary we can conclude that the set of Nash equilibria does not change if we restrict ourselves to behavioral strategies. This is true only in games of perfect recall, and thus, for example, in perfect-information games. We stress again, however, that in general imperfect-information games, mixed and behavioral strategies yield noncomparable sets of equilibria.

5.3 SEQUENTIAL EQUILIBRIUM

We have already seen that the Nash equilibrium concept is too weak for perfect-information games, and how the more selective notion of subgame-perfect equilibrium can be more instructive. The question is whether this essential idea can be applied to the broader class of imperfect-information games; it turns out that it can, although the details are considerably more involved.

Recall that in a subgame-perfect equilibrium we require that the strategy of each agent be a best response in every subgame, not only the whole game. It is immediately apparent that the definition does not apply in imperfect-information games, if for no other reason than we no longer have a well-defined notion of a subgame. What we have instead at each information set is a "subforest" or a collection of subgames. We could require that each player's strategy be a best response in each subgame in each forest, but that would be both too strong a requirement and too weak. To see why it is too strong, consider the game in Figure 5.4.

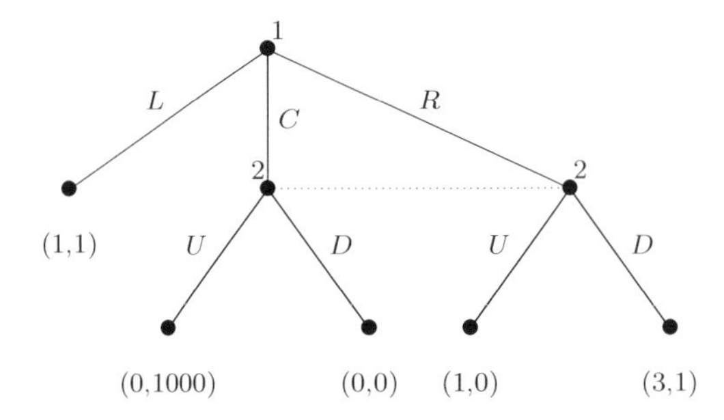

FIGURE 5.4: Player 2 knows where in the information set he is.

The pure strategies of player 1 are $\{L, C, R\}$ and of player 2 $\{U, D\}$. Note also that the two pure Nash equilibria are (L, U) and (R, D). But should either of these be considered "subgame perfect?" On the face of it the answer is ambiguous, since in one subtree U (dramatically) dominates D and in the other D dominates U. However, consider the following argument. R dominates C for player 1, and player 2 knows this. So although player 2 does not have explicit information about which of the two nodes he is in within his information set, he can deduce that he is in the rightmost one based on player 1's incentives, and hence will go D. Furthermore player 1 knows that player 2 can deduce this, and therefore player 1 should go R. Thus, (R, D) is the only subgame-perfect equilibrium.

This example shows how a requirement that a sub-strategy be a best response in all subgames is too simplistic. However, in general it is not the case that subtrees of an information set can be pruned as in the previous example so that all remaining ones agree on the best strategy for the player. In this case the naive application of the SPE intuition would rule out all strategies.

There have been several related proposals that apply the intuition underlying subgame-perfection in more sophisticated ways. One of the more influential notions has been that of *sequential equilibrium* (SE). It shares some features with the notion of trembling-hand perfection, discussed in Section 3.6. Note that indeed trembling-hand perfection, which was defined for normal-form games, applies here just as well; just think of the normal form induced by the extensive-form game. However, this notion makes no reference to the tree structure of the game. SE does, but at the expense of additional complexity.

Sequential equilibrium is defined for games of perfect recall. As we have seen, in such games we can restrict our attention to behavioral strategies. Consider for the moment a fully mixed-strategy profile.[2] Such a strategy profile induces a positive probability on every node in

[2]Again, recall that a strategy is fully mixed if, at every information set, each action is given some positive probability.

the game tree. This means in particular that every information set is given a positive probability. Therefore, for a given fully mixed-strategy profile, one can meaningfully speak of i's expected utility, given that he finds himself in any particular information set. (The expected utility of starting at any node is well defined, and since each node is given positive probability, one can apply Bayes' rule to aggregate the expected utilities of the different nodes in the information set.) If the fully mixed-strategy profile constitutes an equilibrium, it must be that each agent's strategy maximizes his expected utility in each of his information sets, holding the strategies of the other agents fixed.

All of the preceding discussion is for a fully mixed-strategy profile. The problem is that equilibria are rarely fully mixed, and strategy profiles that are not fully mixed do *not* induce a positive probability on every information set. The expected utility of starting in information sets whose probability is zero under the given strategy profile is simply not well defined. This is where the ingenious device of SE comes in. Given any strategy profile S (not necessarily fully mixed), imagine a probability distribution $\mu(h)$ over each information set. μ has to be *consistent* with S, in the sense that for sets whose probability is nonzero under their parents' conditional distribution S, this distribution is precisely the one defined by Bayes' rule. However, for other information sets, it can be any distribution. Intuitively, one can think of these distributions as the new beliefs of the agents, if they are surprised and find themselves in a situation they thought would not occur. This means that agents' expected utility is now well defined in any information set, including those having measure zero. For information set h belonging to agent i, with the associated probability distribution $\mu(h)$, the expected utility under strategy profile S is denoted by $u_i(S \mid h, \mu(h))$.

With this, the precise definition of SE is as follows.

Definition 5.3.1 (Sequential equilibrium). *A strategy profile S is a* sequential equilibrium *of an extensive-form game G if there exist probability distributions $\mu(h)$ for each information set h in G, such that the following two conditions hold:*

1. $(S, \mu) = \lim_{n\to\infty}(S^n, \mu^n)$ *for some sequence $(S^1, \mu^1), (S^2, \mu^2), \ldots$, where S^n is fully mixed, and μ^n is consistent with S^n (in fact, since S^n is fully mixed, μ^n is uniquely determined by S^n); and*
2. *For any information set h belonging to agent i, and any alternative strategy S'_i of i, we have that*

$$u_i(S \mid h, \mu(h)) \geq u_i((S', S_{-i}) \mid h, \mu(h)).$$

Analogous to subgame perfection in games of perfect information, sequential equilibria are guaranteed to always exist.

Theorem 5.3.2. *Every finite game of perfect recall has a sequential equilibrium.*

Finally, while sequential equilibria are defined for games of imperfect information, they are obviously also well defined for the special case of games of perfect information. This raises the question of whether, in the context of games of perfect information, the two solution concepts coincide. The answer is that they almost do, but not quite.

Theorem 5.3.3. *Every subgame-perfect equilibrium is a sequential equilibrium, but the converse is not true in general.*

CHAPTER 6

Repeated and Stochastic Games

In repeated games, a given game (often thought of in normal form) is played multiple times by the same set of players. The game being repeated is called the *stage game*. For example, Figure 6.1 depicts two players playing the Prisoner's Dilemma exactly twice in a row.

This representation of the repeated game, while intuitive, obscures some key factors. Do agents see what the other agents played earlier? Do they remember what they knew? And, while the utility of each stage game is specified, what is the utility of the entire repeated game?

We answer these questions in two steps. We first consider the case in which the game is repeated a finite and commonly-known number of times. Then we consider the case in which the game is repeated infinitely often, or a finite but unknown number of times.

6.1 FINITELY REPEATED GAMES

One way to completely disambiguate the semantics of a finitely repeated game is to specify it as an imperfect-information game in extensive form. Figure 6.2 describes the twice-played Prisoner's Dilemma game in extensive form. Note that it captures the assumption that at each iteration the players do not know what the other player is playing, but afterward they do. Also note that the payoff function of each agent is additive; that is, it is the sum of payoffs in the two-stage games.

The extensive form also makes it clear that the strategy space of the repeated game is much richer than the strategy space in the stage game. Certainly one strategy in the repeated

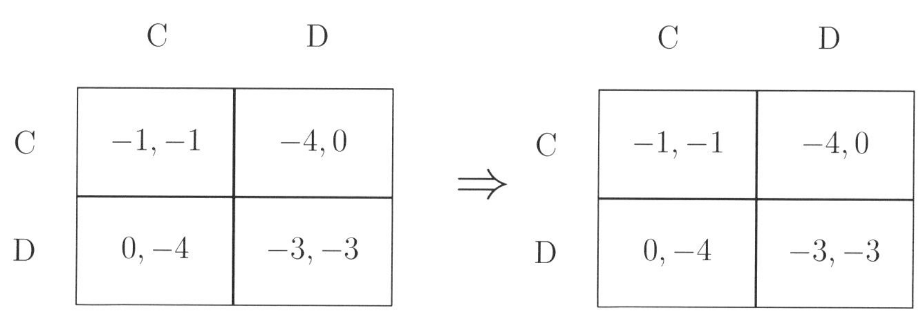

	C	D
C	$-1, -1$	$-4, 0$
D	$0, -4$	$-3, -3$

$\Rightarrow$

	C	D
C	$-1, -1$	$-4, 0$
D	$0, -4$	$-3, -3$

FIGURE 6.1: Twice-played Prisoner's Dilemma.

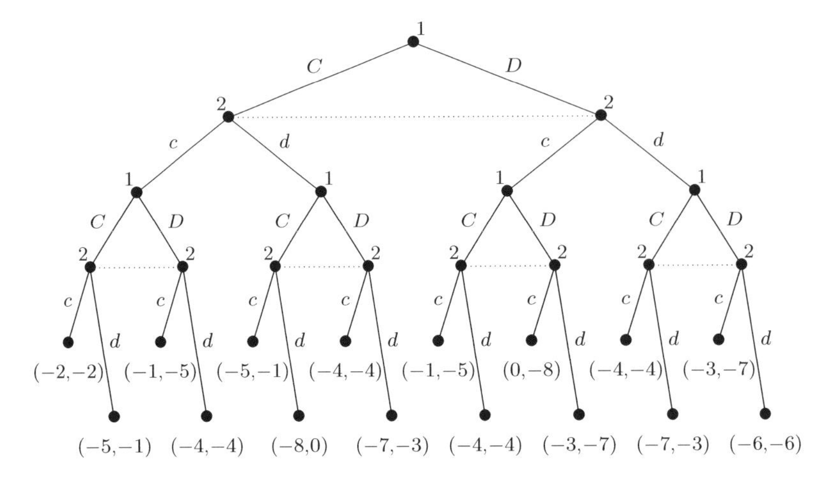

FIGURE 6.2: Twice-played Prisoner's Dilemma in extensive form.

game is to adopt the same strategy in each stage game; clearly, this memory less strategy, called a *stationary strategy*, is a behavioral strategy in the extensive-form representation of the game. But in general, the action (or mixture of actions) played at a stage game can depend on the history of play thus far. Since this fact plays a particularly important role in infinitely repeated games, we postpone further discussion of it to the next section. Indeed, in the finite, known repetition case, we encounter again the phenomenon of backward induction, which we first encountered when we introduced subgame perfect equilibria. Recall that in the Centipede game, discussed in Section 4.3, the unique SPE was to go down and terminate the game at every node. Now consider a finitely repeated Prisoner's Dilemma game. Again, it can be argued, in the last round it is a dominant strategy to defect, no matter what happened so far. This is common knowledge, and no choice of action in the preceding rounds will impact the play in the last round. Thus in the second-to-last round too it is a dominant strategy to defect. Similarly, by induction, it can be argued that the only equilibrium in this case is to always defect. However, as in the case of the Centipede game, this argument is vulnerable to both empirical and theoretical criticisms.

6.2 INFINITELY REPEATED GAMES

When the infinitely repeated game is transformed into extensive form, the result is an infinite tree. So the payoffs cannot be attached to any terminal nodes, nor can they be defined as the sum of the payoffs in the stage games (which in general will be infinite). There are two common ways of defining a player's payoff in an infinitely repeated game to get around this problem. The first is the average payoff of the stage game in the limit.[1]

[1]The observant reader will notice a potential difficulty in this definition, since the limit may not exist. One can extend the definition to cover these cases by using the lim sup operator in Definition 6.2.1 rather than lim.

Definition 6.2.1 (Average reward). *Given an infinite sequence of payoffs* $r_i^{(1)}, r_i^{(2)}, \ldots$ *for player* i*, the* average reward *of* i *is*

$$\lim_{k\to\infty} \frac{\sum_{j=1}^{k} r_i^{(j)}}{k}.$$

The *future discounted reward* to a player at a certain point of the game is the sum of his payoff in the immediate stage game, plus the sum of future rewards discounted by a constant factor. This is a recursive definition, since the future rewards again give a higher weight to early payoffs than to later ones.

Definition 6.2.2 (Discounted reward). *Given an infinite sequence of payoffs* $r_i^{(1)}, r_i^{(2)}, \ldots$ *for player* i*, and a discount factor* β *with* $0 \leq \beta \leq 1$*, the* future discounted reward *of* i *is* $\sum_{j=1}^{\infty} \beta^j r_i^{(j)}$.

The discount factor can be interpreted in two ways. First, it can be taken to represent the fact that the agent cares more about his well-being in the near term than in the long term. Alternatively, it can be assumed that the agent cares about the future just as much as he cares about the present, but with some probability the game will be stopped any given round; $1 - \beta$ represents that probability. The analysis of the game is not affected by which perspective is adopted.

Now let us consider strategy spaces in an infinitely repeated game. In particular, consider the infinitely repeated Prisoner's Dilemma game. As we discussed, there are many strategies other than stationary ones. One of the most famous ones is *Tit-for-Tat*. TfT is the strategy in which the player starts by cooperating and thereafter chooses in round $j + 1$ the action chosen by the other player in round j. Besides being both simple and easy to compute, this strategy is notoriously hard to beat; it was the winner in several repeated Prisoner's Dilemma competitions for computer programs.

Since the space of strategies is so large, a natural question is whether we can characterize all the Nash equilibria of the repeated game. For example, if the discount factor is large enough, both players playing TfT is a Nash equilibrium. But there is an infinite number of others. For example, consider the *trigger strategy*. This is a draconian version of TfT; in the trigger strategy, a player starts by cooperating, but if ever the other player defects then the first defects forever. Again, for sufficiently large discount factor, the trigger strategy forms a Nash equilibrium not only with itself but also with TfT.

The folk theorem—so-called because it was part of the common lore before it was formally written down—helps us understand the space of all Nash equilibria of an infinitely repeated game, by answering a related question. It does not characterize the equilibrium strategy profiles, but rather the payoffs obtained in them. Roughly speaking, it states that in an infinitely repeated game the set of average rewards attainable in equilibrium are precisely those pairs attainable under mixed strategies in a single-stage game, with the constraint on the mixed strategies that each player's payoff is at least the amount he would receive if the other players adopted minmax strategies against him.

More formally, consider any n-player game $G = (N, A, u)$ and any payoff profile $r = (r_1, r_2, \ldots, r_n)$. Let

$$v_i = \min_{s_{-i} \in S_{-i}} \max_{s_i \in S_i} u_i(s_{-i}, s_i).$$

In words, v_i is player i's minmax value: his utility when the other players play minmax strategies against him, and he plays his best response.

Before giving the theorem, we provide some more definitions.

Definition 6.2.3 (Enforceable). *A payoff profile $r = (r_1, r_2, \ldots, r_n)$ is* enforceable *if $\forall i \in N$, $r_i \geq v_i$.*

Definition 6.2.4 (Feasible). *A payoff profile $r = (r_1, r_2, \ldots, r_n)$ is* feasible *if there exist rational, nonnegative values α_a such that for all i, we can express r_i as $\sum_{a \in A} \alpha_a \, u_i(a)$, with $\sum_{a \in A} \alpha_a = 1$.*

In other words, a payoff profile is feasible if it is a convex, rational combination of the outcomes in G.

Theorem 6.2.5 (Folk Theorem). *Consider any n-player normal-form game G and any payoff profile $r = (r_1, r_2, \ldots, r_n)$.*

1. *If r is the payoff profile for any Nash equilibrium s of the infinitely repeated G with average rewards, then for each player i, r_i is enforceable.*
2. *If r is both feasible and enforceable, then r is the payoff profile for some Nash equilibrium of the infinitely repeated G with average rewards.*

Although we do not give the proof of this theorem here, a high-level description of the argument is both instructive and intuitive. The proof proceeds in two parts. The first part uses the definition of minmax and best response to show that an agent can never receive less than his minmax value in any equilibrium. The second part shows how to construct an equilibrium that yields each agent the average payoffs given in any feasible and enforceable payoff profile r. This equilibrium has the agents cycle in perfect lock-step through a sequence of game outcomes that achieve the desired average payoffs. If any agent deviates, the others punish him forever by playing their minmax strategies against him.

Theorem 6.2.5 is actually an instance of a large family of folk theorems. As stated, Theorem 6.2.5 is restricted to infinitely repeated games, to average reward, to the Nash equilibrium, and to games of complete information. However, there are folk theorems that hold for other versions of each of these conditions, as well as other conditions not mentioned here. In particular, there are folk theorems for infinitely repeated games with discounted reward (for a large enough discount factor), for finitely repeated games, for subgame-perfect equilibria (i.e., where agents only administer finite punishments to deviators), and for games of incomplete information. We do not review them here, but the message of each of them is fundamentally the same: the payoffs in the equilibria of a repeated game are essentially constrained only by enforceability and feasibility.

6.3 STOCHASTIC GAMES

Intuitively speaking, a stochastic game is a collection of normal-form games; the agents repeatedly play games from this collection, and the particular game played at any given iteration depends probabilistically on the previous game played and on the actions taken by all agents in that game.

6.3.1 Definition

Definition 6.3.1 (Stochastic game). *A* stochastic game *(also known as a* Markov game*) is a tuple* (Q, N, A, P, R), *where:*

- Q *is a finite set of states;*
- N *is a finite set of* n *players;*
- $A = A_1 \times \cdots \times A_n$, *where* A_i *is a finite set of actions available to player* i;
- $P : Q \times A \times Q \mapsto [0, 1]$ *is the transition probability function;* $P(q, a, \hat{q})$ *is the probability of transitioning from state* s *to state* $\hat{q}$ *after action profile* a*; and*
- $R = r_1, \ldots, r_n$, *where* $r_i : Q \times A \mapsto \mathbb{R}$ *is a real-valued payoff function for player* i.

In this definition we have assumed that the strategy space of the agents is the same in all games, and thus that the difference between the games is only in the payoff function. Removing this assumption adds notation, but otherwise presents no major difficulty or insights. Restricting Q and each A_i to be finite is a substantive restriction, but we do so for a reason; the infinite case raises a number of complications that we wish to avoid.

We have specified the payoff of a player at each stage game (or in each state), but not how these payoffs are aggregated into an overall payoff. To solve this problem, we can use solutions already discussed earlier in connection with infinitely repeated games (Section 6.2). Specifically, the two most commonly used aggregation methods are *average reward* and *future discounted reward*.

Stochastic games are very broad framework, generalizing both Markov decision processes (MDPs) and repeated games. An MDP is simply a stochastic game with only one player, while a repeated game is a stochastic game in which there is only one state (or stage game). Another interesting subclass of stochastic games is zero-sum stochastic games, in which each stage game is a zero-sum game (i.e., for any $q \in Q$, $a \in A$ we have that $\sum_i v_i(q, a) = 0$). Finally, in a *single-controller stochastic game* the transition probabilities depend only on the actions of one particular agent, while players' payoffs still depend on their joint actions.

6.3.2 Strategies and Equilibria

We will now define the strategy space of an agent. Let $h_t = (q^0, a^0, q^1, a^1, \ldots, a^{t-1}, q^t)$ denote a history of t stages of a stochastic game, and let H_t be the set of all possible histories of this length. The set of deterministic strategies is the Cartesian product $\prod_{t, H_t} A_i$, which requires a choice for each possible history at each point in time. As in the previous game forms, an agent's strategy can consist of any mixture over deterministic strategies. However, there are several restricted classes of strategies that are of interest, and they form the following hierarchy. The first restriction is that the mixing take place at each history independently; this is the restriction to behavioral strategies seen in connection with extensive-form games.

Definition 6.3.2 (Behavioral strategy). *A behavioral strategy $s_i(h_t, a_{i_j})$ returns the probability of playing action a_{i_j} for history h_t.*

A Markov strategy further restricts a behavioral strategy so that, for a given time t, the distribution over actions depends only on the current state.

Definition 6.3.3 (Markov strategy). *A* Markov strategy *s_i is a behavioral strategy in which $s_i(h_t, a_{i_j}) = s_i(h'_t, a_{i_j})$ if $q_t = q'_t$, where q_t and q'_t are the final states of h_t and h'_t, respectively.*

The final restriction is to remove the possible dependence on the time t.

Definition 6.3.4 (Stationary strategy). *A* stationary strategy *s_i is a Markov strategy in which $s_i(h_{t_1}, a_{i_j}) = s_i(h'_{t_2}, a_{i_j})$ if $q_{t_1} = q'_{t_2}$, where q_{t_1} and q'_{t_2} are the final states of h_{t_1} and h'_{t_2}, respectively.*

Now we can consider the equilibria of stochastic games, a topic that turns out to be fraught with subtleties. The discounted-reward case is the less problematic case. In this case it can be shown that a Nash equilibrium exists in every stochastic game. In fact, we can state a stronger property. A strategy profile is called a *Markov perfect equilibrium* if it consists of only Markov strategies, and is a Nash equilibrium regardless of the starting state. In a sense, MPE plays a role analogous to the subgame-perfect equilibrium in perfect-information games.

Theorem 6.3.5. *Every n-player, general-sum, discounted-reward stochastic game has a Markov perfect equilibrium.*

The case of average rewards presents greater challenges. For one thing, the limit average may not exist (i.e., although the stage-game payoffs are bounded, their average may cycle and not converge). However, there is a class of stochastic games which is well behaved in this regard. This is the class of *irreducible* stochastic games. A stochastic game is irreducible if every strategy profile gives rise to an irreducible Markov chain over the set of games, meaning that every game can be reached with positive probability regardless of the strategy adopted. In such games the limit averages are well defined, and we have the following theorem.

Theorem 6.3.6. *Every two-player, general-sum, average reward, irreducible stochastic game has a Nash equilibrium.*

Indeed, under the same condition we can state a folk theorem similar to that presented for repeated games in Section 6.2. That is, as long as we give each player an expected payoff that is at least as large as his minmax value, any feasible payoff pair can be achieved in equilibrium through the use of threats.

Theorem 6.3.7. *For every two-player, general-sum, irreducible stochastic game, and every feasible outcome with a payoff vector r that provides to each player at least his minmax value, there exists a Nash equilibrium with a payoff vector r. This is true for games with average rewards, as well as games with large enough discount factors (or, with players that are sufficiently patient).*

CHAPTER 7

Uncertainty About Payoffs: Bayesian Games

All of the game forms discussed so far assumed that all players know what game is being played. Specifically, the number of players, the actions available to each player, and the payoff associated with each action vector, have all been assumed to be common knowledge among the players. Note that this is true even of imperfect-information games; the actual moves of agents are not common knowledge, but the game itself is. In contrast, *Bayesian games*, or games of incomplete information, allow us to represent players' uncertainties about the very game being played.[1] This uncertainty is represented as a probability distribution over a set of possible games. We make two assumptions.

1. All possible games have the same number of agents and the same strategy space for each agent; they differ only in their payoffs.
2. The beliefs of the different agents are posteriors, obtained by conditioning a common prior on individual private signals.

The second assumption is substantive, and we return to it shortly. The first is not particularly restrictive, although at first it might seem to be. One can imagine many other potential types of uncertainty that players might have about the game—how many players are involved, what actions are available to each player, and perhaps other aspects of the situation. It might seem that we have severely limited the discussion by ruling these out. However, it turns out that these other types of uncertainty can be reduced to uncertainty only about payoffs via problem reformulation.

For example, imagine that we want to model a situation in which one player is uncertain about the number of actions available to the other players. We can reduce this uncertainty to uncertainty about payoffs by padding the game with irrelevant actions. For example, consider the following two-player game, in which the row player does not know whether his opponent has only the two strategies L and R or also the third one C:

[1]It is easy to confuse the term "incomplete information" with "imperfect information"; don't...

	L	R
U	1, 1	1, 3
D	0, 5	1, 13

	L	C	R
U	1, 1	0, 2	1, 3
D	0, 5	2, 8	1, 13

Now consider replacing the leftmost, smaller game by a padded version, in which we add a new C column.

	L	C	R
U	1, 1	0, −100	1, 3
D	0, 5	2, −100	1, 13

Clearly the newly added column is dominated by the others and will not participate in any Nash equilibrium (or any other reasonable solution concept). Indeed, there is an isomorphism between Nash equilibria of the original game and the padded one. Thus the uncertainty about the strategy space is reduced to uncertainty about payoffs.

Using similar tactics, it can be shown that it is also possible to reduce uncertainty about other aspects of the game to uncertainty about payoffs only. This is not a mathematical claim, since we have given no mathematical characterization of all the possible forms of uncertainty, but it is the case that such reductions have been shown for all the common forms of uncertainty.

The second assumption about Bayesian games is the *common-prior assumption*. A Bayesian game thus defines not only the uncertainties of agents about the game being played, but also their beliefs about the beliefs of other agents about the game being played, and indeed an entire infinite hierarchy of nested beliefs (the so-called epistemic type space). The common-prior assumption is a substantive assumption that limits the scope of applicability. We nonetheless make this assumption since it allows us to formulate the main ideas in Bayesian games, and without the assumption the subject matter becomes much more involved than is appropriate for this text. Indeed, most (but not all) work in game theory makes this assumption.

7.1 DEFINITION

There are several different ways of presenting Bayesian games; we will offer three definitions of Bayesian games. All three are equivalent, modulo some subtleties which lie outside the scope of this booklet. We include all three since each formulation is useful in different settings, and offers different intuition about the underlying structure of this family of games.

7.1.1 Information Sets

First, we present a definition that is based on information sets. Under this definition, a Bayesian game consists of a set of games that differ only in their payoffs, a common prior defined over them, and a partition structure over the games for each agent.

Definition 7.1.1 (Bayesian game: information sets). *A* Bayesian game *is a tuple* (N, G, P, I) *where:*

- N *is a set of agents;*
- G *is a set of games with* N *agents each such that if* $g, g' \in G$ *then for each agent* $i \in N$ *the strategy space in* g *is identical to the strategy space in* g'*;*
- $P \in \Pi(G)$ *is a common prior over games, where* $\Pi(G)$ *is the set of all probability distributions over* G*; and*
- $I = (I_1, \ldots, I_N)$ *is a tuple of partitions of* G*, one for each agent.*

Figure 7.1 gives an example of a Bayesian game. It consists of four 2×2 games (Matching Pennies, Prisoner's Dilemma, Coordination and Battle of the Sexes), and each agent's partition consists of two equivalence classes.

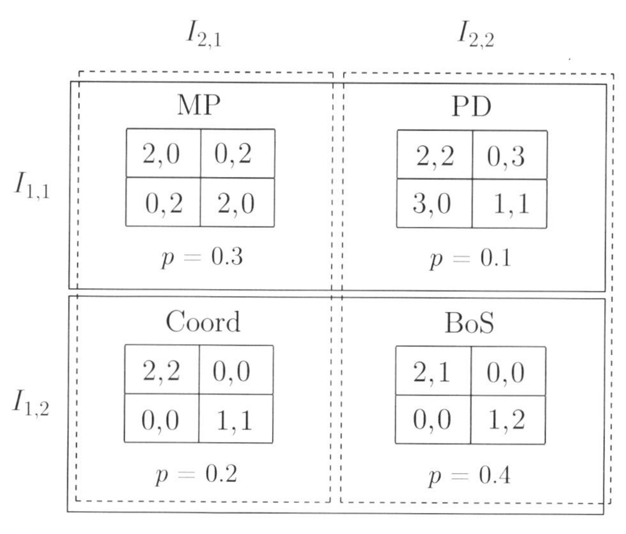

FIGURE 7.1: A Bayesian game.

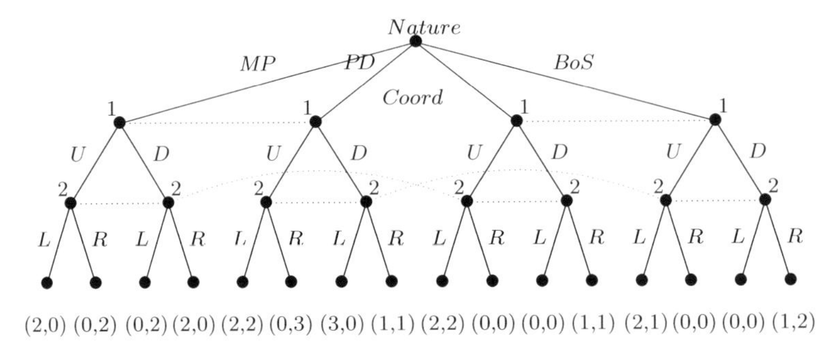

FIGURE 7.2: The Bayesian game from Figure 7.1 in extensive form.

7.1.2 Extensive form with Chance Moves

A second way of capturing the common prior is to hypothesize a special agent called Nature who makes probabilistic choices. While we could have Nature's choice be interspersed arbitrarily with the agents' moves, without loss of generality we assume that Nature makes all its choices at the outset. Nature does not have a utility function (or, alternatively, it can be viewed as having a constant one), and has the unique strategy of randomizing in a commonly known way. The agents receive individual signals about Nature's choice, and these are captured by their information sets in a standard way. The agents have no additional information; in particular, the information sets capture the fact that agents make their choices without knowing the choices of others. Thus, we have reduced games of incomplete information to games of imperfect information, albeit ones with chance moves. These chance moves of Nature require minor adjustments of existing definitions, replacing payoffs by their expectations, given Nature's moves.[2]

For example, the Bayesian game of Figure 7.1 can be represented in extensive form as depicted in Figure 7.2.

Although this second definition of Bayesian games can be initially more intuitive than our first definition, it can also be more cumbersome to work with. This is because we use an extensive-form representation in a setting where players are unable to observe each others' moves. (Indeed, for the same reason we do not routinely use extensive-form games of imperfect information to model simultaneous interactions such as the Prisoner's Dilemma, though we could do so if we wished.) For this reason, we will not make further use of this definition. We close by noting one advantage that it does have, however: it extends very naturally to Bayesian

[2]Note that the special structure of this extensive form game means that we do not have to agonize over the refinements of Nash equilibrium; since agents have no information about prior choices made other than by Nature, all Nash equilibria are also sequential equilibria.

games in which players move sequentially and do (at least sometimes) learn about previous players' moves.

7.1.3 Epistemic Types

Recall that a game may be defined by a set of players, actions, and utility functions. In our first definition agents are uncertain about which game they are playing; however, each possible game has the same sets of actions and players, and so agents are really only uncertain about the game's utility function.

Our third definition uses the notion of an *epistemic type*, or simply a *type*, as a way of defining uncertainty directly over a game's utility function.

Definition 7.1.2 (Bayesian game: types). *A* Bayesian game *is a tuple* (N, A, Θ, p, u) *where:*

- N *is a set of agents;*
- $A = A_1 \times \cdots \times A_n$, *where* A_i *is the set of actions available to player* i*;*
- $\Theta = \Theta_1 \times \ldots \times \Theta_n$, *where* Θ_i *is the type space of player* i*;*
- $p : \Theta \mapsto [0, 1]$ *is a common prior over types; and*
- $u = (u_1, \ldots, u_n)$, *where* $u_i : A \times \Theta \mapsto \mathbb{R}$ *is the utility function for player* i.

The assumption is that all of the above is common knowledge among the players, and that each agent knows his own type. This definition can seem mysterious, because the notion of type can be rather opaque. In general, the type of agent encapsulates all the information possessed by the agent that is not common knowledge. This is often quite simple (e.g., the agent's knowledge of his private payoff function), but can also include his beliefs about other agents' payoffs, about their beliefs about his own payoff, and any other higher-order beliefs.

We can get further insight into the notion of a type by relating it to the formulation at the beginning of this section. Consider again the Bayesian game in Figure 7.1. For each of the agents we have two types, corresponding to his two information sets. Denote player 1's actions as U and D, player 2's actions as L and R. Call the types of the first agent $\theta_{1,1}$ and $\theta_{1,2}$, and those of the second agent $\theta_{2,1}$ and $\theta_{2,2}$. The joint distribution on these types is as follows: $p(\theta_{1,1}, \theta_{2,1}) = .3$, $p(\theta_{1,1}, \theta_{2,2}) = .1$, $p(\theta_{1,2}, \theta_{2,1}) = .2$, $p(\theta_{1,2}, \theta_{2,2}) = .4$. The conditional probabilities for the first player are $p(\theta_{2,1} \mid \theta_{1,1}) = 3/4$, $p(\theta_{2,2} \mid \theta_{1,1}) = 1/4$, $p(\theta_{2,1} \mid \theta_{1,2}) = 1/3$, and $p(\theta_{2,2} \mid \theta_{1,2}) = 2/3$. Both players' utility functions are given in Figure 7.3.

7.2 STRATEGIES AND EQUILIBRIA

Now that we have defined Bayesian games, we must explain how to reason about them. We will do this using the epistemic type definition given earlier, because that is the definition most

a_1	a_2	θ_1	θ_2	u_1	u_2
U	L	$\theta_{1,1}$	$\theta_{2,1}$	2	0
U	L	$\theta_{1,1}$	$\theta_{2,2}$	2	2
U	L	$\theta_{1,2}$	$\theta_{2,1}$	2	2
U	L	$\theta_{1,2}$	$\theta_{2,2}$	2	1
U	R	$\theta_{1,1}$	$\theta_{2,1}$	0	2
U	R	$\theta_{1,1}$	$\theta_{2,2}$	0	3
U	R	$\theta_{1,2}$	$\theta_{2,1}$	0	0
U	R	$\theta_{1,2}$	$\theta_{2,2}$	0	0

a_1	a_2	θ_1	θ_2	u_1	u_2
D	L	$\theta_{1,1}$	$\theta_{2,1}$	0	2
D	L	$\theta_{1,1}$	$\theta_{2,2}$	3	0
D	L	$\theta_{1,2}$	$\theta_{2,1}$	0	0
D	L	$\theta_{1,2}$	$\theta_{2,2}$	0	0
D	R	$\theta_{1,1}$	$\theta_{2,1}$	2	0
D	R	$\theta_{1,1}$	$\theta_{2,2}$	1	1
D	R	$\theta_{1,2}$	$\theta_{2,1}$	1	1
D	R	$\theta_{1,2}$	$\theta_{2,2}$	1	2

FIGURE 7.3: Utility functions u_1 and u_2 for the Bayesian game from Figure 7.1.

commonly used in mechanism design, one of the main applications of Bayesian games. All of the concepts defined below can also be expressed in terms of the first two Bayesian game definitions as well.

The first task is to define an agent's strategy space in a Bayesian game. Recall that in an imperfect-information extensive-form game a pure strategy is a mapping from information sets to actions. The definition is similar in Bayesian games: a pure strategy $\alpha_i : \Theta_i \mapsto A_i$ is a mapping from every type agent i could have to the action he would play if he had that type. We can then define mixed strategies in the natural way as probability distributions over pure strategies. As before, we denote a mixed strategy for i as $s_i \in S_i$, where S_i is the set of all i's mixed strategies. Furthermore, we use the notation $s_j(a_j|\theta_j)$ to denote the probability under mixed strategy s_j that agent j plays action a_j, given that j's type is θ_j.

Next, since we have defined an environment with multiple sources of uncertainty, we will pause to reconsider the definition of an agent's expected utility. In a Bayesian game setting, there are three meaningful notions of expected utility: *ex post*, *ex interim* and *ex ante*. The first is computed based on all agents' actual types, the second considers the setting in which an agent knows his own type but not the types of the other agents, and in the third case the agent does not know anybody's type.

Definition 7.2.1 (*Ex post* expected utility). *Agent i's* ex post *expected utility in a Bayesian game (N, A, Θ, p, u), where the agents' strategies are given by s and the agent' types are given by θ, is defined as*

$$EU_i(s, \theta) = \sum_{a \in A} \left(\prod_{j \in N} s_j(a_j|\theta_j) \right) u_i(a, \theta). \tag{7.1}$$

In this case, the only uncertainty concerns the other agents' mixed strategies, since agent i's *ex post* expected utility is computed based on the other agents' actual types. Of course, in a Bayesian game no agent *will* know the others' types; while that does not prevent us from offering the definition given, it might make the reader question its usefulness. We will see that this notion of expected utility is useful both for defining the other two and also for defining a specialized equilibrium concept.

Definition 7.2.2 (*Ex interim* expected utility). *Agent i's* ex interim *expected utility in a Bayesian game (N, A, Θ, p, u), where i's type is θ_i and where the agents' strategies are given by the mixed-strategy profile s, is defined as*

$$EU_i(s, \theta_i) = \sum_{\theta_{-i} \in \Theta_{-i}} p(\theta_{-i}|\theta_i) \sum_{a \in A} \left(\prod_{j \in N} s_j(a_j|\theta_j) \right) u_i(a, \theta_{-i}, \theta_i). \tag{7.2}$$

or equivalently as

$$EU_i(s, \theta_i) = \sum_{\theta_{-i} \in \Theta_{-i}} p(\theta_{-i}|\theta_i) EU_i(s, (\theta_i, \theta_{-i})). \tag{7.3}$$

Thus, i must consider every assignment of types to the other agents θ_{-i} and every pure action profile a in order to evaluate his utility function $u_i(a, \theta_i, \theta_{-i})$. He must weight this utility value by two amounts: the probability that the other players' types would be θ_{-i} given that his own type is θ_i, and the probability that the pure action profile a would be realized given all players' mixed strategies and types. (Observe that agents' types may be correlated.) Because uncertainty over mixed strategies was already handled in the *ex post* case, we can also write *ex interim* expected utility as a weighted sum of $EU_i(s, \theta)$ terms.

Finally, there is the *ex ante* case, where we compute i's expected utility under the joint mixed strategy s without observing any agents' types.

Definition 7.2.3 (*Ex ante* expected utility). *Agent i's* ex ante *expected utility in a Bayesian game (N, A, Θ, p, u), where the agents' strategies are given by the mixed-strategy profile s, is defined as*

$$EU_i(s) = \sum_{\theta \in \Theta} p(\theta) \sum_{a \in A} \left(\prod_{j \in N} s_j(a_j|\theta_j) \right) u_i(a, \theta) \tag{7.4}$$

or equivalently as

$$EU_i(s) = \sum_{\theta \in \Theta} p(\theta) EU_i(s, \theta) \tag{7.5}$$

or again equivalently as

$$EU_i(s) = \sum_{\theta_i \in \Theta_i} p(\theta_i) EU_i(s, \theta_i). \tag{7.6}$$

Next, we define best response.

Definition 7.2.4 (Best response in a Bayesian game). *The set of agent i's* best responses *to mixed-strategy profile s_{-i} are given by*

$$BR_i(s_{-i}) = \arg\max_{s'_i \in S_i} EU_i(s'_i, s_{-i}). \tag{7.7}$$

Note that BR_i is a set because there may be many strategies for i that yield the same expected utility. It may seem odd that BR is calculated based on i's *ex ante* expected utility. However, write $EU_i(s)$ as $\sum_{\theta_i \in \Theta_i} p(\theta_i) EU_i(s, \theta_i)$ and observe that $EU_i(s'_i, s_{-i}, \theta_i)$ does not depend on strategies that i would play if his type were not θ_i. Thus, we are in fact performing independent maximization of i's *ex interim* expected utility conditioned on each type that he could have. Intuitively speaking, if a certain action is best after the signal is received, it is also the best conditional plan devised ahead of time for what to do should that signal be received.

We are now able to define the Bayes–Nash equilibrium.

Definition 7.2.5 (Bayes–Nash equilibrium). *A* Bayes–Nash equilibrium *is a mixed-strategy profile s that satisfies $\forall i \;\; s_i \in BR_i(s_{-i})$.*

This is exactly the definition we gave for the Nash equilibrium in Definition 2.2.2: each agent plays a best response to the strategies of the other players. The difference from Nash equilibrium, of course, is that the definition of Bayes–Nash equilibrium is built on top of the Bayesian game definitions of best response and expected utility. Observe that we would not be able to define equilibrium in this way if an agent's strategies were not defined for every possible type. In order for a given agent i to play a best response to the other agents $-i$, i must know what strategy each agent would play for each of his possible types. Without this information, it would be impossible to evaluate the term $EU_i(s'_i, s_{-i})$ in Equation (7.7).

7.3 COMPUTING EQUILIBRIA

Despite its similarity to the Nash equilibrium, the Bayes–Nash equilibrium may seem more conceptually complicated. However, as we did with extensive-form games, we can construct a normal-form representation that corresponds to a given Bayesian game.

As with games in extensive form, the induced normal form for Bayesian games has an action for every pure strategy. That is, the actions for an agent i are the distinct mappings from

Θ_i to A_i. Each agent i's payoff given a pure-strategy profile s is his *ex ante* expected utility under s. Then, as it turns out, the Bayes–Nash equilibria of a Bayesian game are precisely the Nash equilibria of its induced normal form. This fact allows us to note that Nash's theorem applies directly to Bayesian games, and hence Bayes–Nash equilibria always exist.

An example will help. Consider again the Bayesian game from Figure 7.3. Note that in this game each agent has four possible pure strategies (two types and two actions). Then player 1's four strategies in the Bayesian game can be labeled UU, UD, DU, and DD: UU means that 1 chooses U regardless of his type, UD that he chooses U when he has type $\theta_{1,1}$ and D when he has type $\theta_{1,2}$, and so forth. Similarly, we can denote the strategies of player 2 in the Bayesian game by RR, RL, LR, and LL.

We now define a 4×4 normal-form game in which these are the four strategies of the two agents, and the payoffs are the expected payoffs in the individual games, given the agents' common prior beliefs. For example, player 2's *ex ante* expected utility under the strategy profile (UU, LL) is calculated as follows:

$$
\begin{aligned}
u_2(UU, LL) &= \sum_{\theta\in\Theta} p(\theta)u_2(U, L, \theta) \\
&= p(\theta_{1,1}, \theta_{2,1})u_2(U, L, \theta_{1,1}, \theta_{2,1}) + p(\theta_{1,1}, \theta_{2,2})u_2(U, L, \theta_{1,1}, \theta_{2,2}) \\
&\quad + p(\theta_{1,2}, \theta_{2,1})u_2(U, L, \theta_{1,2}, \theta_{2,1}) + p(\theta_{1,2}, \theta_{2,2})u_2(U, L, \theta_{1,2}, \theta_{2,2}) \\
&= 0.3(0) + 0.1(2) + 0.2(2) + 0.4(1) = 1.
\end{aligned}
$$

Continuing in this manner, the complete payoff matrix can be constructed as indicated in Figure 7.4.

	LL	LR	RL	RR
UU	2, 1	1, 0.7	1, 1.2	0, 0.9
UD	0.8, 0.2	1, 1.1	0.4, 1	0.6, 1.9
DU	1.5, 1.4	0.5, 1.1	1.7, 0.4	0.7, 0.1
DD	0.3, 0.6	0.5, 1.5	1.1, 0.2	1.3, 1.1

FIGURE 7.4: Induced normal form of the game from Figure 7.3.

	LL	LR	RL	RR
UU	2, 0.5	1.5, 0.75	0.5, 2	0, 2.25
UD	2, 0.5	1.5, 0.75	0.5, 2	0, 2.25
DU	0.75, 1.5	0.25, 1.75	2.25, 0	1.75, 0.25
DD	0.75, 1.5	0.25, 1.75	2.25, 0	1.75, 0.25

FIGURE 7.5: *Ex interim* induced normal-form game, where player 1 observes type $\theta_{1,1}$.

Now the game may be analyzed straightforwardly. For example, we can determine that player 1's best response to RL is DU.

Given a particular signal, the agent can compute the posterior probabilities and recompute the expected utility of any given strategy vector. Thus in the previous example once the row agent gets the signal $\theta_{1,1}$ he can update the expected payoffs and compute the new game shown in Figure 7.5.

Note that for the row player, DU is still a best response to RL; what has changed is how much better it is compared to the other three strategies. In particular, the row player's payoffs are now independent of his choice of which action to take upon observing type $\theta_{1,2}$; in effect, conditional on observing type $\theta_{1,1}$ the player needs only to select a single action U or D. (Thus, we could have written the *ex interim* induced normal form in Figure 7.5 as a table with four columns but only two rows.)

Although we can use this matrix to find best responses for player 1, it turns out to be meaningless to analyze the Nash equilibria in this payoff matrix. This is because these expected payoffs are not common knowledge; if the column player were to condition on his signal, he would arrive at a different set of numbers (though, again, for him best responses would be preserved). Ironically, it is only in the induced normal form, in which the payoffs do not correspond to any *ex interim* assessment of any agent, that the Nash equilibria are meaningful.

Other computational techniques exist for Bayesian games which also have temporal structure—that is, for Bayesian games written using the "extensive form with chance moves" formulation, for which the game tree is smaller than its induced normal form. For example, there is an algorithm for Bayesian games of perfect information that generalizes backward induction (defined in Section 4.4), called *expectimax*. Intuitively, this algorithm is very much like the standard backward induction algorithm given in Figure 4.6. Like that algorithm, expectimax recursively explores a game tree, labeling each non-leaf node h with a payoff vector by examining the labels of each of h's child nodes—the actual payoffs when these child nodes are leaf nodes—and keeping the payoff vector in which the agent who moves at h achieves maximal utility. The new wrinkle is that chance nodes must also receive labels. Expectimax labels a chance node h with a weighted sum of the labels of its child nodes, where the weights are the probabilities that each child node will be selected. This is a popular algorithmic framework for building computer players for perfect-information games of chance such as Backgammon.

7.4 *EX POST* EQUILIBRIA

Finally, working with *ex post* utilities allows us to define an equilibrium concept that is stronger than the Bayes–Nash equilibrium.

Definition 7.4.1 (*Ex post* equilibrium). *An* ex post *equilibrium is a mixed-strategy profile s that satisfies* $\forall\theta$, $\forall i$, $s_i \in \arg\max_{s_i' \in S_i} EU_i(s_i', s_{-i}, \theta)$.

Observe that this definition does not presume that each agent actually *does* know the others' types; instead, it says that no agent would ever want to deviate from his mixed strategy *even if* he knew the complete type vector θ. This form of equilibrium is appealing because it is unaffected by perturbations in the type distribution $p(\theta)$. Said another way, an *ex post* equilibrium does not ever require any agent to believe that the others have accurate beliefs about his own type distribution. (Note that a standard Bayes–Nash equilibrium *can* imply this requirement.) The *ex post* equilibrium is thus similar in flavor to equilibria in dominant strategies, which do not require agents to believe that other agents act rationally.

Indeed, many dominant strategy equilibria are also *ex post* equilibria, making it easy to believe that this relationship always holds. In fact, it does not, as the following example shows. Consider a two-player Bayesian game where each agent has two actions and two corresponding types ($\forall_{i \in N}$, $A_i = \Theta_i = \{H, L\}$) distributed uniformly ($\forall_{i \in N}$, $P(\theta_i = H) = 0.5$), and with the same utility function for each agent i:

$$u_i(a, \theta) = \begin{cases} 10 & a_i = \theta_{-i} = \theta_i; \\ 2 & a_i = \theta_{-i} \neq \theta_i; \\ 0 & \text{otherwise.} \end{cases}$$

In this game, each agent has a dominant strategy of choosing the action that corresponds to his type, $a_i = \theta_i$. An equilibrium in these dominant strategies is not *ex post* because if either agent knew the other's type, he would prefer to deviate to playing the strategy that corresponds to the other agent's type, $a_i = \theta_{-i}$.

Finally, *ex post* equilibria do share another, unfortunate similarity with equilibria in dominant strategies—thcy arc not guaranteed to exist.

CHAPTER 8
Coalitional Game Theory

So far we have concentrated on what has become the dominant branch of game theory, the so-called *noncooperative* variant. We now conclude with an overview of *coalitional game theory*, also known as *cooperative game theory*. As was mentioned at the beginning of Chapter 1, when we introduced noncooperative game theory, the term "cooperative" can be misleading. It does not mean that each agent is agreeable and will follow arbitrary instructions. Rather, it means that the basic modeling unit is the group rather than the individual agent. More precisely, in coalitional game theory we still model the individual preference of agents, but not their possible actions. Instead, we have a coarser model of the capabilities of different groups.

8.1 COALITIONAL GAMES WITH TRANSFERABLE UTILITY

In coalitional game theory our focus is on what groups of agents, rather than individual agents, can achieve. Given a set of agents, a coalitional game defines how well each group (or *coalition*) of agents can do for itself. We are not concerned with how the agents make individual choices within a coalition, how they coordinate, or any other such detail; we simply take the payoff[1] to a coalition as given.

In this chapter we will make the *transferable utility assumption*—that the payoffs to a coalition may be freely redistributed among its members. This assumption is satisfied whenever there is a universal *currency* that is used for exchange in the system, and means that each coalition can be assigned a single value as its payoff.

Definition 8.1.1 (Coalitional game with transferable utility). *A* coalitional game with transferable utility *is a pair* (N, v), *where*

- *N is a finite[2] set of players, indexed by i; and*
- *$v : 2^N \mapsto \mathbb{R}$ associates with each coalition $S \subseteq N$ a real-valued payoff $v(S)$ that the coalition's members can distribute among themselves. The function v is also called the*

[1] Alternatively, one might assign *costs* instead of payoffs to coalitions. Throughout this chapter, we will focus on the case of payoffs; the concepts defined herein can be extended analogously to the case of costs.

[2] Observe that we consider only finite coalitional games. The infinite case is also considered in the literature; many but not all of the results from this chapter also hold in this case.

characteristic function, *and a coalition's payoff is also called its* worth. *We assume that* $v(\emptyset) = 0$.

Ordinarily, coalitional game theory is used to answer two fundamental questions:

1. Which coalition will form?
2. How should that coalition divide its payoff among its members?

It turns out that the answer to (1) is often "the grand coalition"—the name given to the coalition of all the agents in N—though this answer can depend on having made the right choice about (2). Before we go any further in answering these questions, however, we provide a coalitional game example to which we will refer throughout the chapter.

Example 8.1.2 (Voting game). A parliament is made up of four political parties, A, B, C, and D, which have 45, 25, 15, and 15 representatives, respectively. They are to vote on whether to pass a $100 million spending bill and how much of this amount should be controlled by each of the parties. A majority vote, that is, a minimum of 51 votes, is required in order to pass any legislation, and if the bill does not pass then every party gets zero to spend.

More generally, in a voting game, there is a set of agents N and a set of coalitions $\mathcal{W} \subseteq 2^N$ that are *winning* coalitions, that is, coalitions that are sufficient for the passage of the bill if all its members choose to do so. To each coalition $S \in \mathcal{W}$, we assign $v(S) = 1$, and to the others we assign $v(S) = 0$.

8.2 CLASSES OF COALITIONAL GAMES

In this section we will define a few important classes of coalitional games, which have interesting applications as well as useful formal properties. We start with the notion of superadditivity, a property often assumed for coalitional games.

Definition 8.2.1 (Superadditive game). *A game* $G = (N, v)$ *is* superadditive *if for all* $S, T \subset N$, *if* $S \cap T = \emptyset$, *then* $v(S \cup T) \geq v(S) + v(T)$.

Superadditivity is justified when coalitions can always work without interfering with one another; hence, the value of two coalitions will be no less than the sum of their individual values. Note that superadditivity implies that the value of the entire set of players (the "grand coalition") is no less than the sum of the value of any nonoverlapping set of coalitions. In other words, the grand coalition has the highest payoff among all coalitional structures. The voting example we gave earlier is a superadditive game.

Taking noninterference across coalitions to the extreme, when coalitions can never affect one another, either positively or negatively, then we have *additive* (or *inessential*) games.

Definition 8.2.2 (Additive game). *A game $G = (N, v)$ is* additive *(or* inessential*) if for all $S, T \subset N$, if $S \cap T = \emptyset$, then $v(S \cup T) = v(S) + v(T)$.*

A related class of games is that of constant-sum games.

Definition 8.2.3 (Constant-sum game). *A game $G = (N, v)$ is* constant sum *if for all $S \subset N$, $v(S) + v(N \setminus S) = v(N)$.*

Note that every additive game is necessarily constant sum, but not vice versa. As in noncooperative game theory, the most commonly studied constant-sum games are *zero-sum games*.

An important subclass of superadditive games are convex games.

Definition 8.2.4 (Convex game). *A game $G = (N, v)$ is* convex *if for all $S, T \subset N$, $v(S \cup T) \geq v(S) + v(T) - v(S \cap T)$.*

Clearly, convexity is a stronger condition than superadditivity. While convex games may therefore appear to be a very specialized class of coalitional games, these games are actually not so rare in practice. Convex games have a number of useful properties, as we will discuss in the next section.

Finally, we present a class of coalitional games with restrictions on the values that payoffs are allowed to take.

Definition 8.2.5 (Simple game). *A game $G = (N, v)$ is* simple *if for all $S \subset N$, $v(S) \in \{0, 1\}$.*

Simple games are useful for modeling voting situations, such as those described in Example 8.1.2. In simple games we often add the requirement that if a coalition wins, then all larger sets are also winning coalitions (i.e., if $v(S) = 1$, then for all $T \supset S$, $v(T) = 1$). This condition might seem to imply superadditivity, but it does not quite. For example, the condition is met by a voting game in which only 50% of the votes are required to pass a bill, but such a game is not superadditive. Consider two disjoint winning coalitions S and T; when they join to form the coalition $S \cup T$ they do not achieve at least the sum of the values that they achieve separately as superadditivity requires.

When simple games are also constant sum, they are called *proper simple games*. In this case, if S is a winning coalition, then $N \setminus S$ is a losing coalition.

Figure 8.1 graphically depicts the relationship between the different classes of games that we have discussed in this section.

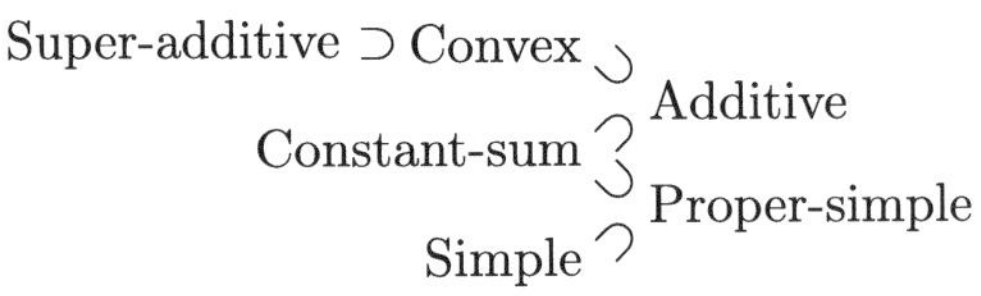

Super-additive $\supset$ Convex $\supset$ Additive

Constant-sum $\supset$ Additive

Constant-sum $\supset$ Proper-simple

Simple $\supset$ Proper-simple

FIGURE 8.1: A hierarchy of coalitional game classes; $X \supset Y$ means that class X is a superclass of class Y.

8.3 ANALYZING COALITIONAL GAMES

The central question in coalitional game theory is the division of the payoff to the grand coalition among the agents. This focus on the grand coalition is justified in two ways. First, since many of the most widely studied games are superadditive, the grand coalition will be the coalition that achieves the highest payoff over all coalitional structures, and hence we can expect it to form. Second, there may be no choice for the agents but to form the grand coalition; for example, public projects are often legally bound to include all participants.

If it is easy to decide to concentrate on the grand coalition, however, it is less easy to decide how this coalition should divide its payoffs. In this section we explore a variety of solution concepts that propose different ways of performing this division.

Before presenting the solution concepts, it is helpful to introduce some terminology. First, let $\psi : \mathbb{N} \times \mathbb{R}^{2^{|N|}} \mapsto \mathbb{R}^{|N|}$ denote a mapping from a coalitional game (that is, a set of agents N and a value function v) to a vector of $|N|$ real values, and let $\psi_i(N, v)$ denote the ith such real value. Denote such a vector of $|N|$ real values as $x \in \mathbb{R}^{|N|}$. Each x_i denotes the share of the grand coalition's payoff that agent $i \in N$ receives. When the coalitional game (N, v) is understood from context, we write x as a shorthand for $\psi(N, v)$.

Now we can give some basic definitions about payoff division.

Definition 8.3.1 (Feasible payoff). *Given a coalitional game* (N, v), *the* feasible payoff set *is defined as* $\{x \in \mathbb{R}^N \mid \sum_{i \in N} x_i \leq v(N)\}$.

In other words, the feasible payoff set contains all payoff vectors that do not distribute more than the worth of the grand coalition.

Definition 8.3.2 (Pre-imputation). *Given a coalitional game* (N, v), *the* pre-imputation set, *denoted* $\mathcal{P}$, *is defined as* $\{x \in \mathbb{R}^N \mid \sum_{i \in N} x_i = v(N)\}$.

We can view the pre-imputation set as the set of feasible payoffs that are *efficient*, that is, they distribute the entire worth of the grand coalition.

Definition 8.3.3 (Imputation). *Given a coalitional game* (N, v), *the* imputation set, $\mathcal{C}$, *is defined as* $\{x \in \mathcal{P} \mid \forall i \in N, x_i \geq v(i)\}$.

Under an imputation, each agent must be guaranteed a payoff of at least the amount that he could achieve by forming a singleton coalition.

Now we are ready to delve into different solution concepts for coalitional games.

8.3.1 The Shapley Value

Perhaps the most straightforward answer to the question of how payoffs should be divided is that the division should be *fair*. Let us begin by laying down axioms that describe what fairness means in our context.

First, say that agents i and j are *interchangeable* if they always contribute the same amount to every coalition of the other agents. That is, for all S that contains neither i nor j, $v(S \cup \{i\}) = v(S \cup \{j\})$. The *symmetry* axiom states that such agents should receive the same payments.

Axiom 8.3.4 (Symmetry). *For any v, if i and j are interchangeable then $\psi_i(N, v) = \psi_j(N, v)$.*

Second, say that an agent i is a *dummy player* if the amount that i contributes to any coalition is exactly the amount that i is able to achieve alone. That is, for all S such that $i \notin S$, $v(S \cup \{i\}) - v(S) = v(\{i\})$. The *dummy player* axiom states that dummy players should receive a payment equal to exactly the amount that they achieve on their own.

Axiom 8.3.5 (Dummy player). *For any v, if i is a dummy player then $\psi_i(N, v) = v(\{i\})$.*

Finally, consider two different coalitional game theory problems, defined by two different characteristic functions v_1 and v_2, involving the same set of agents. The *additivity* axiom states that if we re-model the setting as a single game in which each coalition S achieves a payoff of $v_1(S) + v_2(S)$, the agents' payments in each coalition should be the sum of the payments they would have achieved for that coalition under the two separate games.

Axiom 8.3.6 (Additivity). *For any two v_1 and v_2, we have for any player i that $\psi_i(N, v_1 + v_2) = \psi_i(N, v_1) + \psi_i(N, v_2)$, where the game $(N, v_1 + v_2)$ is defined by $(v_1 + v_2)(S) = v_1(S) + v_2(S)$ for every coalition S.*

If we accept these three axioms, we are led to a strong result: there is always exactly one pre-imputation that satisfies them.

Theorem 8.3.7. *Given a coalitional game (N, v), there is a unique pre-imputation $\phi(N, v) = \phi(N, v)$ that satisfies the Symmetry, Dummy player, Additivity axioms.*

Note that our requirement that $\phi(N, v)$ be a pre-imputation implies that the payoff division be feasible and efficient.

What is this unique payoff division $\phi(N, v)$? It is called the *Shapley value*, and it is defined as follows.

Definition 8.3.8 (Shapley value). *Given a coalitional game* (N, v), *the* Shapley value *of player* i *is given by*

$$\phi_i(N, v) = \frac{1}{N!} \sum_{S \subseteq N \setminus \{i\}} |S|!(|N| - |S| - 1)!\Big[v(S \cup \{i\}) - v(S)\Big].$$

This expression can be viewed as capturing the "average marginal contribution" of agent i, where we average over all the different sequences according to which the grand coalition could be built up from the empty coalition. More specifically, imagine that the coalition is assembled by starting with the empty set and adding one agent at a time, with the agent to be added chosen uniformly at random. Within any such sequence of additions, look at agent i's marginal contribution at the time he is added. If he is added to the set S, his contribution is $[v(S \cup \{i\}) - v(S)]$. Now multiply this quantity by the $|S|!$ different ways the set S could have been formed prior to agent i's addition and by the $(|N| - |S| - 1)!$ different ways the remaining agents could be added afterward. Finally, sum over all possible sets S and obtain an average by dividing by $|N|!$, the number of possible orderings of all the agents.

For a concrete example of the Shapley value in action, consider the voting game given in Example 8.1.2. Recall that the four political parties A, B, C, and D have 45, 25, 15, and 15 representatives, respectively, and a simple majority (51 votes) is required to pass the \$100 million spending bill. If we want to analyze how much money it is fair for each party to demand, we can calculate the Shapley values of the game. Note that every coalition with 51 or more members has a value of \$100 million,[3] and others have \$0. In this game, therefore, the parties B, C, and D are interchangeable, since they add the same value to any coalition. (They add \$100 million to the coalitions $\{B, C\}$, $\{C, D\}$, $\{B, D\}$ that do not include them already and to $\{A\}$; they add \$0 to all other coalitions.) The Shapley value of A is given by:

$$\begin{aligned}
\phi_A &= (3)\frac{(4-1)!(2-1)!}{4!}(100-0) + (3)\frac{(4-3)!(3-1)!}{4!}(100-0) \\
&\quad + (1)\frac{(4-4)!(4-1)!}{4!}(100-100) \\
&= (3)\frac{2}{24}(100) + (3)\frac{(1)(2)}{24}(00-0) + 0 \\
&= 25 + 25 = \$50 \text{ million.}
\end{aligned}$$

[3]Notice that for these calculations we scale the value function to 100 for winning coalitions and 0 for losing coalitions in order to make it align more tightly with our example.

The Shapley value for B (and, by symmetry, also for C and D) is given by:

$$\begin{aligned}\phi_B &= \frac{(4-2)!(2-1)!}{4!}(100-0) + \frac{(4-3)!(3-1)!}{4!}(100-0) \\ &= \frac{2}{24}(100) + \frac{2}{24}(100-0) \\ &= 8.33 + 8.33 = \$16.66 \text{ million.}\end{aligned}$$

Thus the Shapley values are $(50, 16.66, 16.66, 16.66)$, which add up to the entire \$100 million.

8.3.2 The Core

The Shapley value defined a fair way of dividing the grand coalition's payment among its members. However, this analysis ignored questions of stability. We can also ask: would the agents be *willing* to form the grand coalition given the way it will divide payments, or would some of them prefer to form smaller coalitions? Unfortunately, sometimes smaller coalitions can be more attractive for subsets of the agents, even if they lead to lower value overall. Considering the majority voting example, while A does not have a unilateral motivation to vote for a different split, A and B have incentive to defect and divide the \$100 million between themselves (e.g., dividing it $(75, 25)$).

This leads to the question of what payment divisions would make the agents want to form the grand coalition. The answer is that they would want to do so if and only if the payment profile is drawn from a set called the *core*, defined as follows.

Definition 8.3.9 (Core). *A payoff vector x is in the* core *of a coalitional game (N, v) if and only if*

$$\forall S \subseteq N, \sum_{i \in S} x_i \geq v(S).$$

Thus, a payoff is in the core if and only if no sub-coalition has an incentive to break away from the grand coalition share the payoff it is able to obtain independently. That is, it requires that the sum of payoffs to any group of agents $S \subseteq N$ must be at least as large as the amount that these agents could share among themselves if they formed a coalition on their own. Notice that Definition 8.3.9 implies that payoff vectors in the core must always be imputations.

Since the core provides a concept of stability for coalitional games we can see it as an analog of the Nash equilibrium from noncooperative games. However, it is actually a stronger notion: Nash equilibrium describes stability only with respect to deviation by a single agent.

Instead, the core is an analog of the concept of *strong Nash equilibrium*, which requires stability with respect to deviations by arbitrary coalitions of agents.

As a notion of stability for coalitional games, the core is appealing. However, the alert reader might have two lingering doubts, arising due to its implicit definition through inequalities:

1. Is the core always nonempty?
2. Is the core always unique?

Unfortunately, the answer to both questions is no. Let us consider again the Parliament example with the four political parties. The set of minimal coalitions that meet the required 51 votes is $\{A, B\}$, $\{A, C\}$, $\{A, D\}$, and $\{B, C, D\}$. We can see that if the sum of the payoffs to parties B, C, and D is less than \$100 million, then this set of agents has incentive to deviate. On the other hand, if B, C, and D get the entire payoff of \$100 million, then A will receive \$0 and will have incentive to form a coalition with whichever of B, C, and D obtained the smallest payoff. Thus, the core is empty for this game.

On the other hand, when the core is nonempty it may not define a unique payoff vector either. Consider changing our example so that instead of a simple majority, an 80% majority is required for the bill to pass. The minimal winning coalitions are now $\{A, B, C\}$ and $\{A, B, D\}$. Any complete distribution of the \$100 million among parties A and B now belongs to the core, since all winning coalitions must have both the support of these two parties.

These examples call into question the universality of the core as a solution concept for coalitional games. We already saw in the context of noncooperative game theory that solution concepts—notably, the Nash equilibrium—do not yield unique solutions in general. Here we are in an arguably worse situation, in that the solution concept may yield no solution at all.

Luckily, there exist several results that allow us to predict the emptiness or nonemptiness of the core based on a coalitional game's membership in one of the classes we defined in Section 8.2.

Theorem 8.3.10. *Every constant-sum game that is not additive has an empty core.*

We say that a player i is a *veto player* if $v(N \setminus \{i\}) = 0$.

Theorem 8.3.11. *In a simple game the core is empty iff there is no veto player. If there are veto players, the core consists of all payoff vectors in which the nonveto players get zero.*

Theorem 8.3.12. *Every convex game has a nonempty core.*

A final question we consider regards the relationship between the core and the Shapley value. We know that the core may be empty, but if it is not, is the Shapley value guaranteed to lie in the core? The answer in general is no, but the following theorem gives us a sufficient condition for this property to hold. We already know from Theorem 8.3.12 that the core of convex games is nonempty. The following theorem further tells us that for such games the Shapley value belongs to that set.

Theorem 8.3.13. *In every convex game, the Shapley value is in the core.*

History and References

There exist several excellent technical introductory textbooks for game theory, including Osborne and Rubinstein [1994], Fudenberg and Tirole [1991], and Myerson [1991]. The reader interested in gaining deeper insight into game theory should consult not only these, but also the most relevant strands of the the vast literature on game theory which has evolved over the years.

In their seminal book, von Neumann and Morgenstern [1944] introduced the normal-form game, the extensive form, the concepts of pure and mixed strategies, as well as other notions central to game theory and utility theory. Schelling [1960] was one of the first to show that interesting social interactions could usefully be modeled using game theory, for which he was recognized in 2005 with a Nobel Prize.

The literature on Pareto optimality and social optimization dates back to the early twentieth century, including seminal work by Pareto and Pigou, but perhaps was best established by Arrow in his seminal work on social choice [Arrow, 1970]. John Nash introduced the concept of what would become known as the "Nash equilibrium" [Nash, 1950; Nash, 1951], without a doubt the most influential concept in game theory to this date. Indeed, Nash received a Nobel Prize in 1994 because of this work.[1]

In 1928 von Neumann derived the "maximin" solution concept to solve zero-sum normal-form games [von Neumann, 1928]. Nash's proof that all noncooperative games have equilibria [Nash, 1950; Nash, 1951] opened the floodgates to a series of refinements and alternative solution concepts which continues to this day. We covered several of these solution concepts. The minimax regret decision criterion was first proposed by Savage [1954], and further developed in Loomes and Sugden [1982] and Bell [1982]. Recent work from a computer science perspective includes Hyafil and Boutilier [2004], which also applies this criterion to the Bayesian games setting we introduce in Chapter 7. Iterated removal of dominated strategies, and the closely related rationalizability, enjoy a long history, though modern discussion of them is most firmly anchored in two independent and concurrent publications: Pearce [1984] and Bernheim [1984]. Correlated equilibria were introduced in Aumann [1974]; Myerson's quote is taken from Solan and Vohra [2002]. Trembling-hand perfection was introduced in Selten [1975].

[1] John Nash was also the topic of the Oscar-winning 2001 movie *A Beautiful Mind*; however, the movie had little to do with his scientific contributions and indeed got the definition of Nash equilibrium wrong.

The concept of evolutionarily stable strategies (ESSs) again has a long history, but was most explicitly put forward in Maynard Smith and Price [1973]—which also introduced the Hawk–Dove game—and figured prominently a decade later in the seminal Maynard Smith [1982]. Experimental work on learning and the evolution of cooperation appears in Axelrod [1984]. It includes discussion of a celebrated tournament among computer programs that played a finitely repeated Prisoner's Dilemma game and in which the simple Tit-for-Tat strategy emerged victorious.

The earliest game-theoretic publication is arguably that of Zermelo, who in 1913 introduced the notions of a game tree and backward induction and argued that in principle chess admits a trivial solution [Zermelo, 1913]. It was already mentioned earlier that extensive-form games were discussed explicitly in von Neumann and Morgenstern [1944], as was backward induction. Subgame perfection was introduced by Selten [1965], who received a Nobel Prize in 1994. The Centipede game was introduced by Rosenthal [1981]; many other papers discuss the rationality of backward induction in such games [Aumann, 1995; Binmore, 1996; Aumann, 1996].

In 1953 Kuhn introduced extensive-form games of imperfect information, including the distinction and connection between mixed and behavioral strategies [Kuhn, 1953]. Sequential equilibria were introduced by Kreps and Wilson [1982]. Here, as in normal-form games, the full list of alternative solution concepts and connection among them is long, and the interested reader is referred to Hillas and Kohlberg [2002] and Govindan and Wilson [2005] for a more extensive survey than is possible here.

Some of the earliest and most influential work on repeated games is Luce and Raiffa [1957] and Aumann [1959]. Of particular note is that the former provided the main ideas behind the folk theorem and that the latter explored the theoretical differences between finitely and infinitely repeated games. Aumann's work on repeated games led to a Nobel Prize in 2005. Our proof of the folk theorem is based on Osborne and Rubinstein [1994]. Stochastic games were introduced in Shapley [1953]. The state of the art regarding them circa 2003 appears in the edited collection Neyman and Sorin [2003]. Filar and Vrieze [1997] provide a rigorous introduction to the topic, integrating MDPs (or single-agent stochastic games) and two-person stochastic games.

Bayesian games were introduced by Harsanyi [1967–1968]; in 1994 he received a Nobel Prize, largely because of this work.

In the early days of game theory research, coalitional game theory was a major focus, particularly of economists. This is partly because the theory is closely related to equilibrium analysis and seemingly bridges a gap between game theory and economics. Von Neumann and

Morgenstern, for example, devoted more than half of their classic text, *Theory of Games and Economic Behavior* [von Neumann and Morgenstern, 1944], to an analysis of coalitional games. A large body of theoretical work on coalitional game theory has focused on the development of solution concepts, possibly in an attempt to explain the behavior of large systems such as markets. Solid explanations of the many solution concepts and their properties are given by Osborne and Rubinstein [1994] and Peleg and Sudhölter [2003].

References

Arrow, K. J. (1970). *Social choice and individual values.* New Haven, CT: Yale University Press.

Aumann, R. (1959). Acceptable points in general cooperative n-person games. *Contributions to the Theory of Games, 4*, 287–324.

Aumann, R. (1974). Subjectivity and correlation in randomized strategies. *Journal of Mathematical Economics, 1*, 67–96.

Aumann, R. (1995). Backward induction and common knowledge of rationality. *GEB: Games and Economic Behavior, 8*(1), 6–19.

Aumann, R. (1996). Reply to Binmore. *GEB: Games and Economic Behavior, 17*(1), 138–146.

Axelrod, R. (1984). *The evolution of cooperation.* New York: Basic Books.

Bell, D. E. (1982). Regret in decision making under uncertainty. *Operations Research, 30*, 961–981.

Bernheim, B. D. (1984). Rationalizable strategic behavior. *Econometrica, 52*, 1007–1028.

Binmore, K. (1996). A note on backward induction. *GEB: Games and Economic Behavior, 17*(1), 135–137.

Filar, J., and Vrieze, K. (1997). *Competitive Markov decision processes.* Springer-Verlag.

Fudenberg, D., and Tirole, J. (1991). *Game theory.* Cambridge, MA: MIT Press.

Govindan, S., and Wilson, R. (2005). Refinements of Nash equilibrium. In S. Durlauf and L. Blume (Eds.), *The new Palgrave dictionary of economics*, vol. II. New York: Macmillan.

Harsanyi, J. (1967–1968). Games with incomplete information played by "Bayesian" players, parts I, II and III. *Management Science, 14*, 159–182, 320–334, 486–502.

Hillas, J., and Kohlberg, E. (2002). Foundations of strategic equilibrium. In R. Aumann and S. Hart (Eds.), *Handbook of game theory*, vol. III, chapter 42, 1597–1663. Amsterdam: Elsevier.

Hyafil, N., and Boutilier, C. (2004). Regret minimizing equilibria and mechanisms for games with strict type uncertainty. *UAI: Proceedings of the Conference on Uncertainty in Artificial Intelligence* (pp. 268–277).

Kreps, D., and Wilson, R. (1982). Sequential equilibria. *Econometrica, 50*, 863–894.

Kuhn, H. (1953). Extensive games and the problem of information. *Contributions to the Theory of Games* (pp. 193–216). Princeton, NJ: Princeton University Press. Reprinted in H. Kuhn (Ed.), *Classics in Game Theory*, Princeton, NJ: Princeton University Press, 1997.

Loomes, G., and Sugden, R. (1982). Regret theory: An alternative theory of rational choice under uncertainty. *Economic Journal, 92*, 805–824.

Luce, R., and Raiffa, H. (1957). *Games and decisions*. New York: John Wiley and Sons.

Maynard Smith, J. (1982). *Evolution and the theory of games*. Cambridge University Press.

Maynard Smith, J., and Price, G. R. (1973). The logic of animal conflict. *Nature*, *246*, 15–18.

Myerson, R. (1991). *Game theory: Analysis of conflict*. Harvard Press.

Nash, J. (1950). Equilibrium points in n-person games. *Proceedings of the National Academy of Sciences USA*, *36*, 48–49. Reprinted in H. Kuhn (Ed.), *Classics in Game Theory*, Princeton, NJ: Princeton University Press, 1997.

Nash, J. (1951). Non-cooperative games. *Annals of Mathematics*, *54*, 286–295.

Neyman, A., and Sorin, S. (2003). *Stochastic games and applications*. Kluwer Academic Press.

Osborne, M. J., and Rubinstein, A. (1994). *A course in game theory*. Cambridge, MA: MIT Press.

Pearce, D. (1984). Rationalizable strategic behavior and the problem of perfection. *Econometrica*, *52*, 1029–1050.

Peleg, B., and Sudhölter, P. (2003). *Introduction to the theory of cooperative games*. Kluwer Academic Publishers.

Rosenthal, R. (1981). Games of perfect information, predatory pricing and the chain-store paradox. *Journal of Economic Theory*, *25*(1), 92–100.

Savage, L. J. (1954). *The foundations of statistics*. New York: John Wiley and Sons. (2nd edition, Mineola, NY: Dover Press, 1972).

Schelling, T. C. (1960). *The strategy of conflict*. Cambridge, MA: Harvard University Press.

Selten, R. (1965). Spieltheoretische Behandlung eines Oligopolmodells mit Nachfragetraegheit. *Zeitschrift fur die gesamte Staatswissenschaft*, *12*, 301–324.

Selten, R. (1975). Reexamination of the perfectness concept for equilibrium points in extensive games. *International Journal of Game Theory*, *4*, 25–55.

Shapley, L. S. (1953). Stochastic games. *Proceedings of the National Academy of Sciences*, *39*, 1095–1100.

Shoham, Y., and Leyton-Brown, K. (2008). *Multiagent systems: Algorithmic, game-theoretic and logical foundations*. Cambridge, UK: Cambridge University Press.

Solan, E., and Vohra, R. (2002). Correlated equilibrium payoffs and public signalling in absorbing games. *International Journal of Game Theory*, *31*, 91–121.

von Neumann, J. (1928). Zur Theorie der Gesellschaftsspiele. *Mathematische Annalen*, *100*, 295–320.

von Neumann, J., and Morgenstern, O. (1944). *Theory of games and economic behavior*. Princeton, NJ: Princeton University Press.

Zermelo, E. F. F. (1913). Über eine Anwendung der Mengenlehre auf die Theorie des Schachspiels. *Fifth International Congress of Mathematicians*, *II*, 501–504.

Index

ϵ-Nash equilibrium, **27**, 27–28

action, 1, **3**, 4–7, 10–16, 19–21, 23, 25–27, 31–33, 35, 36, 38, 39, 41, 42, 45, 46, 50, 51, 53, 54, 57, 61, 62, 64–67, 69
action profile, **3**, 4, 10, 13, 19, 20, 53, 63
additive game, **71**
alpha-beta pruning, 39
average reward, **51**, 52–55

Backoff game, 2
backward induction, **35**, **38**, 39, 50, 67, 80
Battle of the Sexes game, **6**, 7, 12, 15, 24, 25, 59
Bayes' rule, 47
Bayes–Nash equilibrium, **64**, 65, 67
  ex post, *see ex post* equilibrium
Bayesian game, 3, **57**, 58, **59**, 60, **61**, 62–67, 79, 80
behavioral strategy
  extensive form game, **43**, 44, 45
  stochastic game, **54**
best response, 10, **11**, 13, 14, 16, 18, 19, 21, 23, 24, 27, 28, 30, 34, 36, 44–46, 52, 64, 66
  in a Bayesian game, **64**, 66

Centipede game, **39**, 50, 80
characteristic function, **70**, 73
Church–Rosser property, **22**
coalitional game theory, 1, 69–77, 81
coalitional game with transferable utility, **69**
common-payoff game, **4**, 5, 10
common-prior assumption, **58**
constant-sum game
  coalitional, **71**, 76
  noncooperative, **5**
convex game, **71**, 76, 77
Coordination game, 5, 59
core, **75**, 76, 77
correlated equilibrium, **24, 25**, 26

descendant, 32, 37
dominant solvable, *see* solvable by iterated elimination
dominant strategy, 20, 21, 43, 44, 50, 67, 68
dominated strategy, 20, **21**, 22, 24, 28, 79
dummy player, 5, **73**

efficiency
  coalitional, 72, 73
empirical frequency, **14**
epistemic type, 58, **61**
equilibrium, *see* solution concept
equilibrium in dominant strategies, **20**
evolutionarily stable strategy, 28–30
  weak, 29
ex post equilibrium, 67
expected utility, **7**, 8, 16, 17, 25, 47, 62–64, 66
  ex ante, **63**, 64, 65
  ex interim, **63**, 64
  ex post, **62**, 63
expectimax algorithm, **67**

extensive form game, 31–48, 80
  imperfect information, **41**, 41–48, 60
  perfect information, **32**, 31–41, 80
  with chance nodes, 60–61, 67

feasible payoff, 55, **72**
fixed-point theorem, **13**
folk theorem, 51, **52**, 55, 80
fully mixed strategy, **7**, 26, 27, 46, 47
future discounted reward, **51**, 53

game in matrix form, *see* normal form game
game in strategic form, *see* normal form game
game of incomplete information, *see* Bayesian game
game tree, *see* extensive form game, perfect information

Hawk–Dove game, 29, 80
history, 32, 50, 54

imperfect-information game, *see* extensive form game, imperfect information
imputation, **72**, 73, 75
induced normal form, 31, 64–67
inessential game, 71
information set
  Bayesian game, 59–62
  extensive form game, 41, 42, 44–47
interchangeable agents, **73**
irreducible stochastic game, **55**

Markov decision problem (MDP), 53, 80
Markov game, **53**
Markov perfect equilibrium (MPE), **54**
Markov strategy, **54**
Matching Pennies game, **5**, 6, 13, 17, 18, 23, 59
matrix form game, *see* normal form game
maximum regret, **19**, 20
maxmin strategy, **15**, 16–19
maxmin value, **15**, 16–18
mechanism design, **21**, 62
minimax algorithm, **39**
minimax regret, 18, **19**, 20, 79
minmax strategy, 15, **16**, 17, 18, 52
minmax value, **16**, 17, 18, 52, 55
mixed strategy, **7**, 8, 11–15, 17, 20–22, 24, 26, 28–30, 34, 35, 43–45, 52, 62, 63, 67, 79
  support of, **7**, 11, 12, 14, 30
mixed-strategy profile, **7**, 8, 16, 27, 46, 47, 63, 64, 67

Nash equilibrium, **11**, 9–76
  ϵ-, *see* ϵ-Nash equilibrium
  Bayes–, *see* Bayes–Nash equilibrium
  strict, **11**, 13
  weak, **11**, 12
normal form game, **3**, 3–9, 15, 31, 34, 35, 43, 46, 49, 64–67

optimal strategy, **9**

Pareto domination, **10**, 20
Pareto optimality, 9, **10**, 20, 79
payoff function, **3**, 49, 53, 61
perfect equilibrium, *see* trembling-hand perfect equilibrium
perfect recall, **44**, 45, 46, 48
perfect-information game, *see* extensive form game, perfect information

pre-imputation, **72**, 73
Prisoner's Dilemma game, **2**, 3–4, 6, 20, 22, 23, 35, 42, 43, 49–51, 59, 60, 80
proper equilibrium, **27**
proper simple game, **71**
pure coordination game, **4**, 5
pure strategy, **6**, 7, 10–14, 17, 21, 23, 24, 26, 32, 33, 35, 42–44, 46, 62, 64, 65
pure-strategy profile, **6**, 7, 65

rationalizable strategy, **23, 24**
regret, **19**
repeated game, 14, 16, 49–53, 55, 80
Rochambeau game, *see* Rock, Paper, Scissors game
Rock, Paper, Scissors game, 5, **6**

security level, **15**
sequential equilibrium, **46**, 45–48
Shapley value, **74**, 75–77
Sharing game, 32, 33
simple game, **71**, 76
single-controller stochastic game, **54**
solution concept, **9**, 10, 11, **15**, 16, 20, 21, 24, 26, 27, 31, 48, 58, 72, 73, 76, 79–81
  coalitional
    core, *see* core
    Shapley value, *see* Shapley value
  noncooperative
    ϵ-Nash equilibrium, *see* ϵ-Nash equilibrium
    Bayes–Nash equilibrium, *see* Bayes–Nash equilibrium
    correlated equilibrium, *see* correlated equilibrium
    dominant solvable, *see* dominant solvable
    equilibrium in dominant strategies, *see* equilibrium in dominant strategies
    maxmin strategies, *see* maxmin strategy
    minmax strategies, *see* minmax strategy
    Pareto optimality, *see* Pareto optimality
    perfect equilibrium, *see* trembling-hand perfect equilibrium
    proper equilibrium, *see* proper equilibrium
    rationalizable strategies, *see* rationalizable strategy
    sequential equilibrium, *see* sequential equilibrium
    strong Nash equilibrium, *see* strong Nash equilibrium
    subgame-perfect equilibrium, *see* subgame-perfect equilibrium
solvable by iterated elimination, 22
stage game, **49**, 50, 51, 53
stationary strategy, **50**, **54**
stochastic game, **53**, 54, 55, 80
strategic form game, *see* normal form game
strategy
  behavioral, *see* behavioral strategy
  Markov, *see* Markov strategy
  mixed, *see* mixed strategy
  pure, *see* pure strategy
  stationary, *see* stationary strategy
  trigger, *see* trigger strategy
strict domination, **20**, 22
strict Pareto efficiency, **10**, *see* Pareto optimality
strong Nash equilibrium, 76
subgame, 31, 35–38, 45–48, 50, 53, 54, 80

subgame-perfect equilibrium (SPE), 31, 35, **37**, 38, 45, 46, 48, 53, 54
superadditive game, **70**, 71
support, *see* mixed strategy, support of

TCP user's game, **2**
team games, **4**
tit-for-tat (TfT), **51**, 80
traffic, 4
transferable utility assumption, **69**
tree-form game, *see* extensive form game
trembling-hand perfect equilibrium, 26, **27**
trigger strategy, **51**
type, 58, 61–68

utility function, **1**, **3**, 9, 32, 39, 60–62, 67
utility theory, **1**, 4, 79

value of a zero-sum game, **17**
very weak domination, **20**, 23
veto player, **76**

weak domination, **20**, 23

zero-sum game, **5**, 10, 17, 39, 54, **71**

Printed in Great Britain
by Amazon